COLLINS

Nations
OF THE
WORLD
ATLAS

COLLINS

Nations
OF THE
WORLD
ATLAS

HarperCollins*Publishers*

Collins Nations of the World Atlas

Collins
An Imprint of HarperCollins*Publishers*
77-85 Fulham Palace Road, Hammersmith, London W6 8JB

First published as Times Guide to the Nations of the World 1994
First published as Collins Nations of the World Atlas 1996.

Printed by The Edinburgh Press Ltd.

ISBN 0 00 448367 7

JH8544

Editorial research and text:
Geographical Research Associates, Maidenhead

The maps in this product are also available for purchase in digital format,
from Bartholomew Data Sales,
Tel: +44 (0)181 307 4065, Fax: +44 (0)181 307 4813

MAP LEGEND & ABBREVIATIONS 6

NATIONS OF THE WORLD *(map)* 6

TIME ZONES *(map)* 7

WORLD ECONOMIC GROUPINGS *(map)* 8

INTERNATIONAL ORGANIZATIONS *(map)* 9

GROSS NATIONAL PRODUCT *(map)* 10

POPULATION DENSITY *(map)* 11

NATIONS OF THE WORLD 12-219

ANTARCTICA *(map)* 220

INDEX 221-3

NATIONS AND CONTINENTS OF THE WORLD

To help the reader identify the geographical location of each nation the maps have been colour coded in the following way;

- Africa
- Asia
- Australasia
- Europe
- North America
- South America

MAP LEGEND FOR ATLAS SECTION

——	International boundary	▲	Mountain height (in metres)
----	Disputed or undefined international boundary	🝙	Lake
——	Main road	🝙	Seasonal lake or salt pan
——	Main railway	✈	Major airport
——	River	■	Capital city
······	Canal	•	Main city or town
----	Seasonal river		

ABBREVIATIONS

The following abbreviations have been used. Codes given in brackets following the name of a currency are those issued by the International Standards Organization.

ANZUS	Australia, New Zealand, United States Security Treaty
ASEAN	Association of Southeast Asian Nations
CARICOM	Caribbean Community
CACM	Central American Common Market
CIS	Commonwealth of Independent States
Col. Plan	Colombo Plan
Comm.	Commonwealth
ECOWAS	Economic Community of West African States
EEA	European Economic Area
EFTA	European Free Trade Association
EU	European Union
G7	Group of seven industrial nations (Canada, France, Germany, Italy, Japan, UK, USA)
Mercosur	Common Market of the Southern Cone
NAFTA	North American Free Trade Area
NATO	North Atlantic Treaty Organization
OAS	Organization of American States
OAU	Organization of African Unity
OECD	Organization for Economic Co-operation and Development
OPEC	Organization of Petroleum Exporting Countries
OSCE	Organization for Security and Co-operation in Europe
SADC	Southern African Development Community
UN	United Nations
WEU	Western European Union
GDP	Gross Domestic Product
GNP	Gross National Product

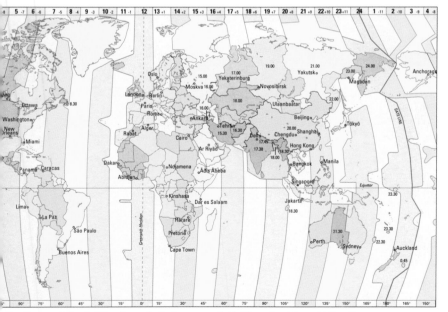

Zone Times are the Standard Times kept on land and sea compared with 12 hours (noon) Greenwich Mean Time. Daylight Saving Time (normally one hour in advance of local Standard Time), which is observed by certain countries for part of the year, is not shown on the map.

INTERNATIONAL ORGANIZATIONS (Right)

- Council of Europe
- Commonwealth of Independent States
- Organization of African Unity (OAU)
- Arab League
- Organization of American States (OAS)
- Commonwealth
- Not a member of any of the organizations shown on the map

THE UNITED NATIONS

The United Nations is the largest international group of countries. It was formed in 1945 in order to promote world peace and co-operation between nations. Its headquarters are in New York. Here the 185 members regularly meet in a General Assembly to settle disputes and agree on common policies to world problems. The work of the United Nations is carried out through its various agencies which include:

Agency:	Responsibility:
UNESCO	Science, education and culture.
UNICEF	Children's welfare.
UNDRO	Disaster relief.
UNHCR	Aid to refugees.
WHO	Health.
FAO	Food & agriculture.
UNEP	Environment.
UNDP	Development programme.

Note:- Countries represented by colour stripes are those which belong to more than one of the Economic Groups shown on the map.

WORLD ECONOMIC GROUPINGS

Note:- Countries represented by colour stripes are those which are members of more than one of the International Organizations shown on the map.

ECONOMIC GROUPS (Left)

North American Free Trade Area (NAFTA)

Common Market of the Southern Cone (Mercosur)

Caribbean Community (CARICOM)

Central American Common Market (CACM)

Economic Community of West African States (ECOWAS)

Organization for Economic Co-operation and Development (OECD)

Colombo Plan

Organization for Petroleum Exporting Countries (OPEC)

Economic Community of Central African States (CEEAC)

Southern African Development Community (SADC)

Association of Southeast Asian Nations (ASEAN)

Not a member of any of the organizations shown on the map

European Union (EU)

European Economic Area (EEA)

European Free Trade Association (EFTA)

FASTEST GROWING POPULATIONS
(average % per annum 1990-1995)

	Country	Growth
1	Afghanistan	5.83
2	Andorra	5.50
3	Yemen	4.97
4	Jordan	4.89
5	French Guiana	4.53

HIGHEST POPULATIONS 1994

	Country	Population
1	China	1 208 841 000
2	India	918 570 000
3	USA	260 650 000
4	Indonesia	192 217 000
5	Brazil	153 725 000

LOWEST LIFE EXPECTANCY
(life expectancy at birth, in years) 1990-95 figs

	Country	Men	Country	Women
1	Sierra Leone	37.5	Sierra Leone	40.6
2	Guinea Bissau	41.9	Afghanistan	44.0
3	Afghanistan	43.0	Guinea	45.0
4	Gambia	43.4	Guinea Bissau	45.1
5	Uganda	43.6	Malawi Uganda	46.2

GROSS NATIONAL PRODUCT

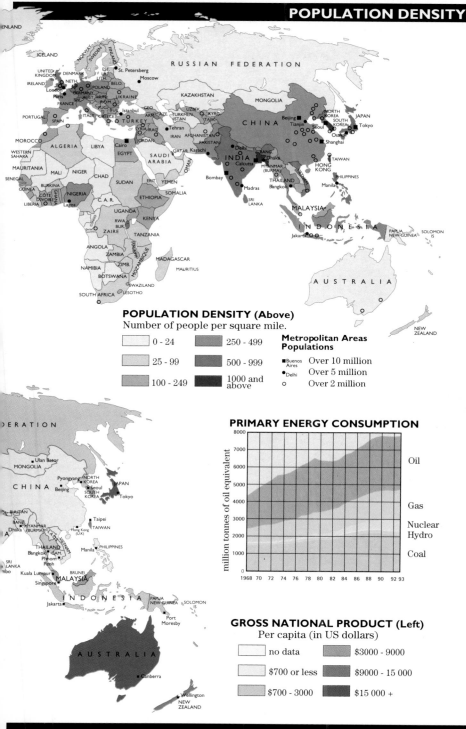

POPULATION DENSITY (Above)
Number of people per square mile.

0 - 24	250 - 499
25 - 99	500 - 999
100 - 249	1000 and above

Metropolitan Areas Populations

■ Buenos Aires — Over 10 million
● Delhi — Over 5 million
○ — Over 2 million

PRIMARY ENERGY CONSUMPTION

million tonnes of oil equivalent

8000
7000
6000
5000
4000
3000
2000
1000
0

1968 70 72 74 76 78 80 82 84 86 88 90 92 93

Oil

Gas

Nuclear
Hydro

Coal

GROSS NATIONAL PRODUCT (Left)
Per capita (in US dollars)

no data	$3000 - 9000
$700 or less	$9000 - 15 000
$700 - 3000	$15 000 +

AFGHANISTAN

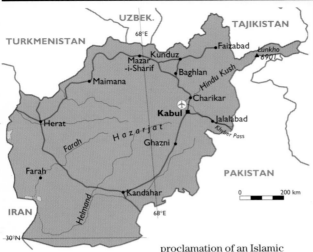

PHYSICAL GEOGRAPHY

The country is predominantly mountainous with the highest range, the Hindu Kush, rising to over 7,500 m (24,600 ft). The ranges are bordered by fertile alluvial plains. In the south there is a desert plateau.

CLIMATE

Afghanistan has a climate of extremes – summers are hot and dry and the winters are cold with heavy snowfalls, particularly in the northern mountains. Annual rainfall varies between 101–406 mm (4–16 inches) a year.

POLITICS AND RECENT HISTORY

A republic was established in 1973 when a military coup ended the 53-year monarchy. In 1978 a further coup introduced a pro-Soviet government. A year later an uprising against the government provoked a Soviet invasion, followed by ten years of Soviet occupation. Fourteen years of civil war between Mujahedeen factions and Communists officially ended in April 1992 with the fall of President Najibullah's Soviet-backed regime and proclamation of an Islamic republic (which has since been designated a state).

Although the Mujahedeen opposition movements were united during the civil war against the common enemy, warring broke out between the various factions as the Soviet troops pulled out. Kabul, relatively unscathed during the civil war, had been a haven for refugees, its population trebling. In 1992 fierce fighting broke out among the Mujahedeen for control of Kabul, causing thousands of civilian deaths.

Fighting has continued since then, mainly between the government forces of President Burhanuddin Rabbani and troops loyal to Gulbuddin Hakmatyar who, under the terms of an earlier agreement, was to have become prime minister but was prevented from entering the city to take up his post. More recently, a third force with fundamental Islamic leanings, known as the Taliban Student Militia and led by a mullah Muhammed Umar, has posed a significant military threat. United Nations (UN) attempts to secure peace have not met with success and President Rabbani has declared himself unwilling to step down and make way for a multi-party council, as envisaged in the UN plan. Nevertheless, there are signs that government forces are slowly gaining ground and bringing a measure of stability to some provinces. Certainly, large numbers of refugees are now returning from Iran and Pakistan.

ECONOMY

The economy is based on agriculture. One-third of farms were abandoned during the civil war, but agricultural land is now being reclaimed by a UN mine-clearing agency.

Illegal selling of opium is rife and the country has been declared the world's leading supplier by the UN Drug Control Programme. The main legitimate exports are fruit, nuts, wool, skins, carpets and cotton.

THE STATE OF AFGHANISTAN	
STATUS	Republic
AREA	652,225 sq km (251,773 sq miles)
POPULATION	17,691,000
CAPITAL	Kabul
LANGUAGE	Pushtu, Dari
RELIGION	90% Sunni, 9% Shi'a Muslim. Hindu, Sikh and Jewish minorities
CURRENCY	Afghani (AFA)
ORGANIZATIONS	Col. Plan, UN

PHYSICAL GEOGRAPHY
The country is rugged and mountainous, with coastal plains. The highest point, Mt. Korab, reaches 2,751 m (9,025 ft).

CLIMATE
The climate has warm summers and cool winters, more severe in the mountains. Rain falls mainly between October and May, at around 1,353–1,425 mm (54–57 inches) per year.

POLITICS AND RECENT HISTORY
For 46 years Albania existed under an extreme Communist regime. The country was internationally isolated and controlled by secret police. When Sali Berisha became Albania's first non-Communist president in 1992, he inherited the poorest country in Europe, with half the population unemployed, and a backward rural economy. With an end to Communism and isolation, Albania joined the Council for Security and Co-operation in Europe, the International Monetary Fund and the World Bank, and now wishes to join the North Atlantic Treaty Organization.

Albania faces conflict near its southern border in an area with a Christian Greek minority, known by the Greeks as Northern Epirus. Albanians fear that Greece has territorial designs on this area, whereas Greece believes Albania is persecuting the ethnic Greek minority. Following the collapse of Communism, thousands of illegal immigrants crossed from Albania into Greece, in search of farm work. However, many thousands have been deported by Greece at times of tension between the two countries.

Now with a Muslim majority (having had 23 years of official atheism under Communist rule), Albania has close ties with Turkey and has joined the Islamic Conference Organization.

Albania is receiving military equipment and training advice from the USA and, in return, US soldiers of the United Nations team stationed in former Yugoslavia have been permitted access through Albania in order to monitor Macedonia.

ECONOMY
Between 1990 and 1993 the Albanian economy virtually collapsed and industrial output fell by 60 per cent as industrial complexes – which employed thousands in the Communist era – shut down. One-tenth of the Albanian population emigrated. Since then, a new economy has emerged. Much of the former state-owned land was split up into small plots under a new privatization law, and agricultural output is now rising by over 14 per cent a year. Small businesses are developing and the first foreign investment was evident in a Coca Cola bottling plant. Tirana's new hotel and business complex has financial support from the European Bank for Reconstruction and Development. The World Bank is contributing towards new roads and agricultural development.

Albania has some natural resources, notably chrome, copper, oil and gas. The country's principal sources of income are foreign aid and remittances from Albanians working abroad. These remittances have become Albania's largest source of foreign exchange.

THE REPUBLIC OF ALBANIA	
STATUS	Republic
AREA	28,750 sq km (11,100 sq miles)
POPULATION	3,389,000
CAPITAL	Tirana (Tiranë)
LANGUAGE	Albanian (Gheg, Tosk)
RELIGION	70% Muslim, 20% Greek Orthodox, 10% Roman Catholic
CURRENCY	lek (ALL)
ORGANIZATIONS	UN

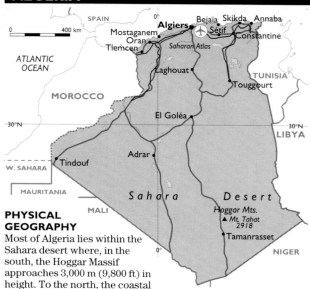

PHYSICAL GEOGRAPHY

Most of Algeria lies within the Sahara desert where, in the south, the Hoggar Massif approaches 3,000 m (9,800 ft) in height. To the north, the coastal plain is fringed by the Atlas mountains.

CLIMATE

Most of Algeria is hot, with negligible rainfall. Along the Mediterranean coast temperatures are more moderate, with most rain during the mild winters. Temperatures range between 12–34°C (52–93°F) over the year and rainfall varies between 158–762 mm (6–30 inches), depending on altitude.

POLITICS AND RECENT HISTORY

In 1989 a new constitution, approved by referendum, allowed for multi-party elections, thus offering an end to many years of one-party socialist rule. However, elections due in January 1992 were cancelled when they promised the return of an Islamic fundamentalist regime led by the Islamic Socialist Front (FIS). President Bendjedid resigned and a five-member High Council of State, with strong army representation, took power. This administration has grown in unpopularity, facing increasing violence from Islamic groups who have adopted guerrilla tactics in order to achieve their objectives. Terrorist attacks and assassinations have been countered by ever more repressive measures, and the civil conflict now claims at least 25,000 lives each year. Algeria's president, General Lamine Zeroual (who succeeded the five-man collective presidency in 1994) has sought dialogue with the FIS but so far these attempts have come to nothing. Although Madani Merzak, the commander of the FIS military wing, has recently shown a more tolerant attitude, the avowed aim of multi-party democracy seems far away.

THE DEMOCRATIC AND POPULAR REPUBLIC OF ALGERIA	
STATUS	Republic
AREA	2,381,745 sq km (919,355 sq miles)
POPULATION	26,722,000
CAPITAL	Algiers (El-Djezair)
LANGUAGE	83% Arabic, French, Berber
RELIGION	Muslim
CURRENCY	Algerian dinar (DZD)
ORGANIZATIONS	Arab League, OAU, OPEC, UN

ECONOMY

The economy is built on agriculture, manufacturing and the exploitation of oil, gas and mineral resources. However, the government has failed to develop this wealth, a factor which contributes significantly to the political instability. Some reform has begun: foreign companies are being allowed into the mining industry and, with World Bank assistance, giant state-owned complexes are being transformed into small, more diverse companies.

The main agricultural crops are cereals, vines, olives, citrus fruits, dates and tomatoes. Phosphates are manufactured at a fertilizer complex at Annaba, which is to be expanded. Other minerals exploited are iron ore, lead, zinc and mercury. The production of oil has declined in recent years, but natural gas output has greatly increased.

The Maghreb-Europe pipeline, running from Algeria through Morocco to Córdoba in Spain, will allow a greater volume of liquefied natural gas exports to Europe. The murder of Europeans working on the pipeline, however, illustrates Algeria's dilemma. Economic well-being is unachievable while political turbulence remains.

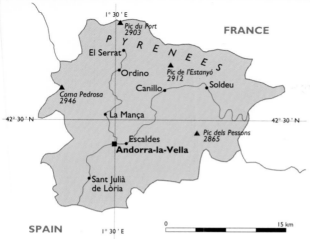

PHYSICAL GEOGRAPHY
Andorra is a small mountainous state in the Pyrenees, and is entirely 1,800 m (5,900 ft) above sea-level. It consists of deep valleys and gorges.

CLIMATE
The climate is alpine. The long winter, which lasts for six months, is severe with heavy snowfall. Spring, however, is mild and summer warm. Annual temperatures range between 2–19°C (36–37°F). Rainfall averages 808 mm (32 inches) over the year.

POLITICS AND RECENT HISTORY
Andorra is a co-principality whose status dates back to the 13th century, the 'princes' being the President of France and the Bishop of Urgel in Spain. The co-principality had ruled itself in semi-feudal fashion for seven centuries, with France conducting foreign affairs. Andorra is now an independent constitutional monarchy, the co-princes sharing only the titular role of Head of State, or Monarch.
    In 1993 Andorra acquired a new constitution, approved earlier in the year by referendum, which allows for an elected parliament with political parties, a judicial system and control of foreign policy, although consultation is still necessary on matters affecting France or Spain.
    A feature of the principality is the small size of the electorate. The proportion of foreigners in the population is high, and only the native 20 per cent are eligible to vote.
    In the principality's first constitutional election, held in 1993, five political groups competed for the 28 seats in the Andorran parliament, known as the General Council of the Valleys. Oscar Ribas Reig, heading the centre-right Andorran National Democrat party, became head of a coalition government. In late 1994 the government resigned, after losing a parliamentary vote of confidence, and Ribas Reig was replaced by Marc Forné Molné of the Liberal Union, heading a minority government.

ECONOMY
Each year 12 million visitors are attracted to Andorra by duty-free shopping, where electronics and designer clothing are the main markets. In winter it is an important skiing centre. As much as 80 per cent of the work force is engaged in tourism, and this is the major industry.
    Other income is derived from the sale of electric power to Catalonia, postage stamps, livestock and tobacco. Financial services are on the increase. There is no income tax in Andorra and the per capita income is among the highest in the world.
    Andorra has signed a treaty to join the European Union (EU) customs union, allowing it the same privileges when selling goods within the EU as if it were a full member. In 1993, the country became a member of the United Nations in its own right.

THE PRINCIPALITY OF ANDORRA	
STATUS	**Principality**
AREA	**465 sq km (180 sq miles)**
POPULATION	**65,000**
CAPITAL	**Andorra la Vella**
LANGUAGE	**Catalan, Spanish, French**
RELIGION	**Roman Catholic majority**
CURRENCY	**French franc (FRF) Spanish peseta (ESP)**
ORGANIZATIONS	**Council of Europe, UN**

PHYSICAL GEOGRAPHY

Plateaux generally exceeding 1,000 m (3,300 ft) predominate. There is a pronounced central highland area around Huambo where the highest point reaches 2,100 m (6,900 ft).

CLIMATE

The climate shows great diversity and ranges from desert in the south to hot, humid equatorial conditions in the north. Rainfall varies from 609 mm (24 inches) in the north to 254 mm (10 inches) in the south. The cold Benguela current exerts a moderating influence along the coastline.

POLITICS AND RECENT HISTORY

Angola achieved independence from Portugal in 1975, whereupon civil war broke out. Initially, the rival factions were the Popular Movement for the Liberation of Angola (MPLA), originally supported by the Soviet Union and Cuba, the Front for the Liberation of Angola (FNLA) which enjoyed the backing of the USA, and the National Union for the Total Liberation of Angola (UNITA) whose leader, Dr Jonas Savimbi, was supported by South Africa. In 1984 the FNLA surrendered to the MPLA and the US government switched its allegiance to UNITA.

Sixteen years of conflict eventually ended with a peace accord in 1991, jointly engineered by the USA, Russia and Portugal. South African and Cuban forces left the country, the MPLA renounced one-party rule and multi-party elections were duly held in September 1992. They attracted a large turn-out and were generally held to be fair by United Nations (UN) and international observers. President José Eduardo dos Santos won about 49 per cent and Dr Savimbi around 40 per cent of the vote. However, Savimbi, unable to accept defeat, resumed the civil war which became more vicious and was estimated to claim 1,000 lives per day, mostly through starvation. Although UNITA reached a position where it held 70 per cent of the country and besieged government-held towns, Savimbi forfeited all international support and forced agreement to peace negotiations, through UN pressure. The UN has so far achieved some success with a cease-fire, separation of the warring factions and the introduction of a peace-keeping force. Dr Savimbi now appears to have submitted to the dos Santos presidency.

ECONOMY

Angola is a fertile land and possesses considerable wealth in the form of diamonds, oil, iron ore and other minerals. The government receives revenue amounting to $3 billion from diamonds and oil, but despite this, people throughout the country have suffered starvation. Massive inflation has rendered the currency worthless and aid has been ineffective because of corruption. However, now that there are prospects for peace, the economy should improve.

THE REPUBLIC OF ANGOLA	
STATUS	Republic
AREA	1,246,700 sq km (481,225 sq miles)
POPULATION	10,276,000
CAPITAL	Luanda
LANGUAGE	Portuguese, tribal dialects
RELIGION	mainly traditional beliefs, Roman Catholic and Protestant minorities
CURRENCY	new kwanza (AOK)
ORGANIZATIONS	OAU, SADC, UN

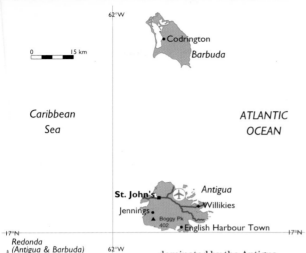

PHYSICAL GEOGRAPHY

With a land area of 280 sq km (108 sq miles), Antigua is low and undulating with hills rising to a 402 m (1,320 ft) peak in the southwest. The coast is indented with many bays fringed with coral reefs. Barbuda is a low, wooded coral island covering an area of 161 sq km (62 sq miles), while a third island, Redonda, 1 sq km (0.3 sq miles), is no more than an uninhabited rocky outcrop.

CLIMATE

The climate is tropical, although modified by the trade winds and sea breezes. Rainfall is around 1,050 mm (42 inches) a year, low for the region, and comes mainly in the summer. Temperatures average 27°C (81°F) but can rise to 33°C (92°F) between May and November.

GOVERNMENT AND RECENT HISTORY

Antigua and Barbuda became fully independent as a constitutional monarchy within the British Commonwealth in 1981. The British monarch is head of state, represented locally by a governor-general. The political scene has been dominated by the Antigua Labour Party (ALP), led for nearly 45 years by Vere C. Bird, first as chief minister and then prime minister. The domination of the Labour Party continued with their victory in the March 1994 general election, when they won 11 of the 17 parliamentary seats. The opposition United Progressive Party, which had one seat in the previous parliament, took five in the 1994 election. Following the victory, Vere C. Bird resigned as prime minister, at the age of 84, and was succeeded by his son Lester Bird.

ECONOMY

The main source of revenue is tourism, which rose to prominence particularly in the five years between 1985 and 1990. During this time, it contributed towards a steady 6 per cent economic growth and now currently accounts for 60 per cent of gross domestic product. The country is seeking diversification to lessen the reliance on tourism, and sea island cotton, previously the country's main export, is making a comeback.

Agriculture is encouraged to lighten the dependency on food imports, but development is restricted by the lack of surface water sources and adverse weather conditions. Lobster fishing is an important activity on Barbuda.

Industry consists of an oil refinery and light industries such as the production of rum, clothing, household appliances and electronics assembly. Puerto Rico is an important export market.

Other sources of revenue include rent from two US military bases and, more recently, financial services. There is, however, a large trade deficit and overseas aid is required.

ANTIGUA AND BARBUDA	
STATUS	Commonwealth State
AREA	442 sq km (171 sq miles)
POPULATION	65,962
CAPITAL	St John's (on Antigua)
LANGUAGE	English
RELIGION	Anglican Christian majority
CURRENCY	E Caribbean dollar (XCD)
ORGANIZATIONS	Caricom, Comm., OAS, UN

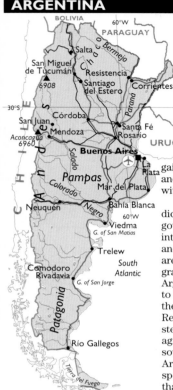

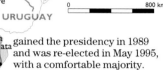

PHYSICAL GEOGRAPHY
Relief is highest in the west in the Andes mountains, where altitudes exceed 6,000 m (19,500 ft). East of the Andes there are fertile plains known as the Pampas.

CLIMATE
In the northern rainforests of the Chaco, hot tropical conditions exist. Central Argentina lies in temperate latitudes, but the southernmost regions are cold, wet and stormy.

POLITICS AND RECENT HISTORY
Civilian government returned to Argentina in 1983, when a constitutional federal democracy replaced years of autocratic military rule. Carlos Menem of the Justicialist Party, founded by General Perón,

gained the presidency in 1989 and was re-elected in May 1995, with a comfortable majority.

Since the demise of military dictatorship, Argentinian governments have gained international respectability, and suspicions that the military are not fully under control are gradually being allayed. Argentina has contributed forces to United Nations operations in the Gulf, Bosnia and Cyprus. Relations with the UK have steadily improved following agreement to disagree about sovereignty of the Falklands. Argentina has proposed a special status for the islands – that they should form part of, but not be fully integrated with, the state of Argentina, which Britain inevitably rejected. More recently, it has been reported that the Argentine Government will pay Falkland Islanders large sums of money (up to $1.5 million) to renounce their British citizenship.

ECONOMY
Agricultural products still account for some 40 per cent of export revenue, although grain crops are affected by falling prices and beef exports by strong competition from western Europe. Industry, including petrochemicals, steel, cars and food processing, is now the chief export earner. There are oil and gas reserves and an abundant supply of hydro-electric power.

Under Menem's presidency inflation, which in 1989 stood at 200 per cent per month, has been crushed and each year the economy has grown. This dramatic improvement has been achieved by a programme of deregulation, privatization, elimination of subsidies and reductions in import tariffs. A crucial step, convertibility, has involved fixing the exchange rate rigidly at one peso to one US dollar. This measure has achieved stability and attracted substantial inward investment but the peso is now overvalued, imports have soared, exports have declined and unemployment has increased. The government has, therefore, adopted strict fiscal measures.

In 1991 Argentina joined Brazil, Uruguay and Paraguay in forming Mercosur (the Common Market of the Southern Cone) which came into effect on 1 January 1995.

THE ARGENTINE REPUBLIC	
STATUS	Republic
AREA	2,766,889 sq km (1,068,302 sq miles)
POPULATION	34,180,000
CAPITAL	Buenos Aires
LANGUAGE	Spanish
RELIGION	90% Roman Catholic, 2% Protestant, Jewish minority
CURRENCY	peso (ARP)
ORGANIZATIONS	Mercosur, OAS, UN

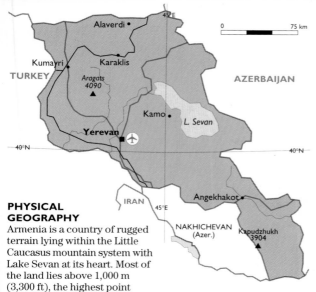

PHYSICAL GEOGRAPHY

Armenia is a country of rugged terrain lying within the Little Caucasus mountain system with Lake Sevan at its heart. Most of the land lies above 1,000 m (3,300 ft), the highest point reaching 4,000 m (13,100 ft).

CLIMATE

The climate, much influenced by altitude, shows marked seasonal variation. Summers are warm and winters cold. Rainfall, although occurring throughout the year, is heaviest in summer.

POLITICS AND RECENT HISTORY

President Levan Ter-Petrosian was returned to power in Armenia, newly independent from the former USSR, when he won 83 per cent of the vote in the October 1991 elections. His government, which has established fairly good relations with Russia, enjoys reasonable stability. The dominant political issue has been the prolonged war over Nagorno-Karabakh, an enclave of Armenian Orthodox Christians within the territory of Azerbaijan. Armenia always denied direct intervention, claiming that the war was waged by natives of Nagorno-Karabakh against the Azeris. However, during 1994 a cease-fire was agreed between

the two nations and so far both parties have avoided any further outbreak of serious conflict. Permanent peace is likely to require significant US aid, based on renunciation of war. President Ter-Petrosian's administration is, however, losing international support as allegations of authoritarianism, press censorship and the suppression of political opponents surface.

ECONOMY

Although Armenia achieved some military success in the war over Nagorno-Karabakh, the nation has suffered economic privation because of the conflict. Azerbaijan has been

able to exert a stranglehold on Armenia, particularly in energy supplies. The last gas pipeline supplying Armenia through Georgia was blown up in August 1993. Although a nuclear power station, the first in the Trans-Caucasus, has been opened, the country suffers from acute energy shortages.

Armenia has few natural resources, but does exploit building materials and some mineral ores. Agriculture is dependent upon irrigation and the main crops are vegetables, particularly sugar-beet and potatoes, fruit and tobacco. Industrial activity contributes nearly 70 per cent of the economy but is nevertheless generally small-scale.

The government has sought to develop a market economy and has pioneered land reform – eighty per cent of the land had been returned to private ownership by mid-1992. Further privatization is planned and foreign investment is being encouraged. Tax reforms have been introduced.

Before the break-up of the Soviet Union, Armenia had developed as a centre of the computer industry. Now, with western support, a fledgling software industry is emerging.

THE REPUBLIC OF ARMENIA	
STATUS	Republic
AREA	30,000 sq km (11,580 sq miles)
POPULATION	3,732,000
CAPITAL	Yerevan
LANGUAGE	Armenian, Russian
RELIGION	Russian Orthodox, Armenian Catholic
CURRENCY	dram
ORGANIZATIONS	CIS, UN

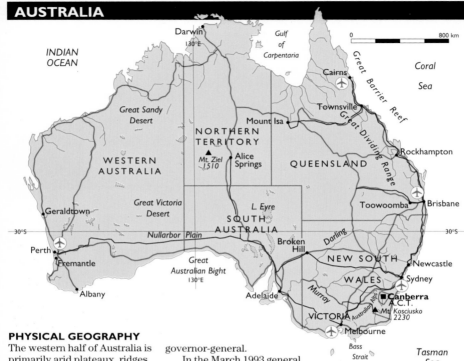

Darwin
130°E
Gulf
of
Carpentaria
INDIAN
OCEAN
Coral
Sea
Cairns
Great Barrier Reef
Townsville
Great Sandy
Desert
Mount Isa
NORTHERN
TERRITORY
Rockhampton
Great Dividing Range
WESTERN
AUSTRALIA
Mt. Ziel
1510
Alice
Springs
QUEENSLAND
Great Victoria
Desert
L. Eyre
Toowoomba
Brisbane
Geraldtown
SOUTH
AUSTRALIA
30°S
Nullarbor Plain
Broken
Hill
Darling
30°S
Perth
Fremantle
Great
Australian Bight
130°E
NEW SOUTH
Newcastle
WALES
Sydney
Albany
Adelaide
Murray
Canberra
A.C.T.
Australian Alps
Mt. Kosciusko
2230
VICTORIA
Melbourne
Bass
Strait
Tasman
Sea
Tasmania
Hobart
0 — 800 km

PHYSICAL GEOGRAPHY

The western half of Australia is primarily arid plateaux, ridges and vast deserts. The central-eastern area comprises lowlands of river systems which drain into Lake Eyre, while to the east is the Great Dividing Range, its highest peak Mt. Kosciusko reaching 2,230 m (7,300 ft).

CLIMATE

Over the continent, the climate varies from cool temperate to tropical monsoon. Rainfall is high only in the northeast, where it exceeds 1,000 mm (39 inches) annually, and decreases markedly from the coast to the interior which is hot and dry. Over 50 per cent of the land area comprises desert and scrub with less than 250 mm (10 inches) of rain a year.

POLITICS AND RECENT HISTORY

The Commonwealth of Australia was founded in 1901. The British Monarch is head of state, represented by a governor-general.

In the March 1993 general election the Australian Labor Party (ALP) retained control (contrary to opinion polls), defeating the conservative Liberal/National Party coalition.

However, in March 1996 the Liberal Party, under its leader John Howard, defeated the incumbent Labor Party under Paul Keating, ending 13 years of Labor rule.

During his administration Keating championed several issues which gained both him and his party a substantial

following. He advocated a change in the constitutional status of the country, proposing that Australia should become a republic within the Commonwealth, because of increasing links with Asia and lessening ties with Europe. It still seems possible that a majority of the population will support constitutional change and that Australia will become a republic by 2001.

A second issue has concerned Aboriginal claims to historic ownership of land. In 1992 a High Court ruling overturned a pre-existing legal assumption that Australia was uninhabited prior to European settlement in the 18th century. The ruling set a precedent for further claims by the Aborigines, creating

THE COMMONWEALTH OF AUSTRALIA	
STATUS	**Federal Nation**
AREA	**7,682,300 sq km (2,965,370 sq miles)**
POPULATION	**17,803,000**
CAPITAL	**Canberra**
LANGUAGE	**English**
RELIGION	**75% Christian, Aboriginal beliefs, Jewish minority**
CURRENCY	**Australian dollar (AUD)**
ORGANIZATIONS	**ANZUS, Col. Plan, Comm., OECD, UN**

controversy in the farming and mining sectors. Progress in this direction stalled somewhat, with the then opposition arguing that money would be better spent on health and education, and the matter was referred to a Select Committee.

ECONOMY

Australia is rich in both agricultural and natural resources. The country is the world's leading producer of wool, which together with wheat, meat, sugar and dairy products account for over 40 per cent of export revenue. There are vast reserves of coal, oil, natural gas, nickel, iron ore, bauxite and uranium ores. Gold, silver, lead, zinc and copper ores are also exploited. Minerals now account for over 30 per cent of Australia's export revenue.

New areas of commerce have been created in eastern Asia, particularly in Japan, to counteract the sharp decline of the traditional European markets. Recent high deficits in the balance of trade have been caused by fluctuation in world demand and competition from the European Union. Foreign debts total A$160 billion and imports are increasing.

However, the Australian economy is growing again as it emerges from recession. Wool exports are improving and sugar production is expanding, despite a severe drought in Queensland in 1994. Unemployment is declining, although interest rates have remained high to guard against inflation. Tourism is now Australia's biggest revenue earner. It showed massive growth during the 1980s and, although the rate of increase has slowed, the Olympic Games, to be held in Sydney in 2000, will provide an important boost – the event is expected to attract an additional 1.5 million visitors.

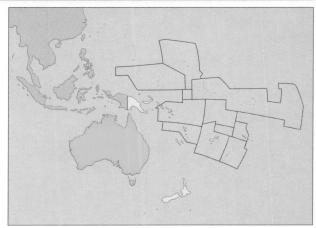

AUSTRALIAN CAPITAL TERRITORY	
STATUS	Federal Territory
AREA	2,432 sq km (939 sq miles)
POPULATION	294,000
CAPITAL	Canberra

SOUTH AUSTRALIA	
STATUS	State
AREA	984,380 sq km (79,970 sq miles)
POPULATION	1,456,000
CAPITAL	Adelaide

NEW SOUTH WALES	
STATUS	State
AREA	801,430 sq km (309,350 sq miles)
POPULATION	5,959,000
CAPITAL	Sydney

TASMANIA	
STATUS	State
AREA	68,330 sq km (26,375 sq miles)
POPULATION	470,000
CAPITAL	Hobart

NORTHERN TERRITORY	
STATUS	Territory
AREA	1,346,200 sq km (519,635 sq miles)
POPULATION	167,000
CAPITAL	Darwin

VICTORIA	
STATUS	State
AREA	227,600 sq km (87,855 sq miles)
POPULATION	4,449,000
CAPITAL	Melbourne

QUEENSLAND	
STATUS	State
AREA	1,727,000 sq km (666,620 sq miles)
POPULATION	3,031,000
CAPITAL	Brisbane

WESTERN AUSTRALIA	
STATUS	State
AREA	2,525,500 sq km (974,845 sq miles)
POPULATION	1,657,000
CAPITAL	Perth

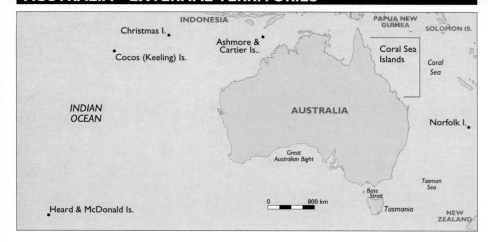

TERRITORY OF ASHMORE & CARTIER ISLANDS

STATUS............**External Territory of Australia**

AREA ....**3 sq km (1.2 sq miles)**

POPULATION.............**no permanent population**

CAPITAL..............................**none**

COCOS (KEELING) ISLANDS

STATUS............**External Territory of Australia**

AREA..............................**14 sq km (5 sq miles)**

POPULATION **647**

CAPITAL..................**Bantam Village**

HEARD AND McDONALD ISLANDS

STATUS............**External Territory of Australia**

AREA ....**412 sq km (159 sq miles)**

POPULATION.............**no permanent population**

CAPITAL........................**Edmonton**

CHRISTMAS ISLAND

STATUS............**External Territory of Australia**

AREA ....**135 sq km (52 sq miles)**

POPULATION **1,275**

CAPITAL.................**Flying Fish Cove**

TERRITORY OF CORAL SEA ISLANDS

STATUS............**External Territory of Australia**

AREA..............................**22 sq km (8.5 sq miles)**

POPULATION.............**no permanent population**

CAPITAL..............................**none**

NORFOLK ISLAND

STATUS............**External Territory of Australia**

AREA..............................**36 sq km (14 sq miles)**

POPULATION **1,977**

CAPITAL ....**Kingston**

PHYSICAL GEOGRAPHY

Austria is an alpine, land-locked country, characterized by mountains, valleys and lakes. To the east are low hills around the valley of the Danube.

CLIMATE

The climate is subject to variation according to altitude, but in general the summers are warm while average winter temperatures are around freezing. Rain falls throughout the year, often as snow in winter, but most occurs in the summer months between May and August.

POLITICS AND RECENT HISTORY

Austria is a Presidential Federal Republic. In the elections of July 1992 Thomas Klestil secured the presidency, replacing the former United Nations Secretary General Kurt Waldheim. This change signalled an improvement in Austria's external relations; during Waldheim's tenure the nation suffered some diplomatic isolation because of his alleged association with Nazi war crimes.

The Austrian government is a coalition of long-standing between the Social Democratic Party (SPO) and the Austrian People's Party (OVP). Franz Vranitzhy (SPO) is chancellor and Erhard Busek (OVP) is vice-chancellor. In elections held in late 1994 the coalition suffered losses, mainly to the right wing Freedom Party, but was able to retain power.

On 1 January 1995 Austria fulfilled its ambition to join the European Union (EU), following a decisive vote in a referendum six months earlier. Negotiations towards membership had been difficult – Austria required satisfactory arrangements to cover products which had components or origins in eastern Europe and also looked for consideration of its alpine farmers.

Early 1995 also saw Austria joining NATO's Partnership for Peace, signifying possible abandonment of its long standing neutrality.

ECONOMY

The Austrian economy is based on industry and includes iron and steel, chemicals and transport equipment, with Germany taking over a third of exports. Tourism, in both summer and winter, is an important foreign exchange earner.

Austria has been affected by recession in Europe, particularly in Germany, although the worst effects have been tempered by a growing economic relationship with the emerging nations of eastern Europe. Some stringent spending cuts have nevertheless been necessary.

Within a week of joining the EU, Austria also joined the Union's exchange rate mechanism, revealing a determination to be at the centre of economic and monetary union within the EU. Austrian membership also provides a convenient bridge between other members of the Union and countries of eastern Europe, particularly Hungary, where Austria has become involved in various joint ventures. Austria has been developing similar initiatives in Slovenia, Slovakia and the Czech Republic.

THE REPUBLIC OF AUSTRIA	
STATUS	Federal Republic
AREA	83,855 sq km (32,370 sq miles)
POPULATION	8,015,000
CAPITAL	Vienna (Wien)
LANGUAGE	German
RELIGION	89% Roman Catholic, 6% Protestant
CURRENCY	schilling (ATS)
ORGANIZATIONS	Council of Europe, EEA, EU, OECD, UN

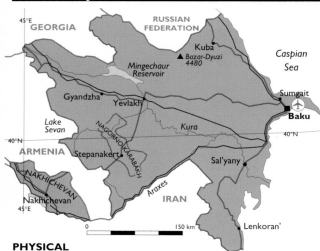

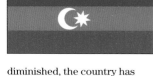

PHYSICAL GEOGRAPHY

Western parts of the country, which include the territory of Nakhichevan west of Armenia, lie within the mountainous regions of the Little Caucasus. In the northeast, the southern limits of the Caucasus Range extend into Azerbaijan, although much of the land adjoining the Caspian Sea is flat and low-lying.

CLIMATE

The climate is continental with hot summers and cold winters, especially in the highlands. Rainfall is light overall with the heaviest falls in summer.

POLITICS AND RECENT HISTORY

Azerbaijan became formally independent from the former Soviet Union in 1991. In September of that year, Ayaz Mutalibov received over 80 per cent of the vote in presidential elections, but he resigned six months later following public protests. In the subsequent elections of June 1992 Abulfaz Elchibey was elected, but was ousted a year later by a military-led coup. His successor, Haydar Aliyev, was confirmed as president at elections in October 1993, although these were thought to be less democratic than those held previously.

A major reason for the downfall of the constitutional leadership was the continued military failure in the conflict with Armenia over the disputed territory of Nagorno-Karabakh. A cease-fire was reached in early 1994, with Moscow's help and after Azerbaijan had rejoined the Commonwealth of Independent States.

Although Azerbaijan's quarrel with Armenia has diminished, the country has been beset by internal strife with several attempted coups, all of which have so far been resisted by President Aliyev.

ECONOMY

Warfare and political instability have taken their toll on the Azerbaijani economy. Since independence, industrial and food production have declined substantially. Unemployment has risen and the economy operates well below capacity.

Despite this, Azerbaijan is a comparatively wealthy state, having major oil reserves beneath and around the Caspian Sea. The government has negotiated with an international consortium for the exploitation of two new oilfields in the Caspian, which are likely to earn significant revenue for many years. However, the Russian government is apt to hinder the exploitation of its southern neighbour's oil wealth.

Successive governments have taken preliminary steps towards developing a market economy, and the country has reached a trade agreement with Turkey, from whom it has received support in its current economic difficulties.

THE REPUBLIC OF AZERBAIJAN	
STATUS	Republic
AREA	87,000 sq km (33,580 sq miles)
POPULATION	7,391,000
CAPITAL	Baku
LANGUAGE	83% Azeri, 6% Armenian, 6% Russian
RELIGION	83% Muslim, Armenian Apostolic, Orthodox
CURRENCY	manat
ORGANIZATIONS	CIS, UN

NAKHICHEVAN	
STATUS	Autonomous Republic of Azerbaijan
AREA	5,500 sq km (2,120 sq miles)
POPULATION	300,000
CAPITAL	Nakhichevan

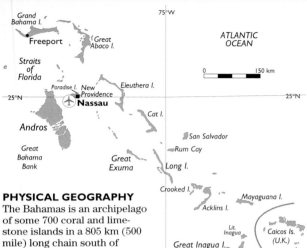

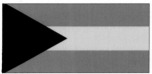

PHYSICAL GEOGRAPHY
The Bahamas is an archipelago of some 700 coral and limestone islands in a 805 km (500 mile) long chain south of Florida. Only 22 islands are inhabited, and over half the population live on New Providence.

CLIMATE
The climate is generally warm throughout the year although the proximity of the North American continent can lead to spells of cool weather in winter. Rainfall is plentiful; the summer months are wettest and the islands are occasionally struck by hurricanes.

POLITICS AND RECENT HISTORY
The Commonwealth of the Bahamas, which became independent from Britain in 1973, is a parliamentary democracy based on the Westminster model. Elections to the House of Assembly are for a five-year term. In August 1992 the Free National Movement won victory over the Progressive Liberal Party and Hubert Ingraham replaced the long-serving Sir Lynden Pindling as prime minister. Accusations of government involvement with drug trafficking have influenced political activity in recent years. Pressure from the USA has forced the introduction of counter-measures.

The Bahamian government has had to introduce strict measures to deter the settling of illegal immigrants from Haiti and Jamaica

ECONOMY
The Bahamian economy is heavily dependent on tourism, which contributes about 70 per cent of GDP. Around 90 per cent of visitors are from the USA. Between 1988-91 the industry showed a decline attributed to weakening in the North American economies and competition from other Caribbean resorts although signs of improvement followed a 10 per cent rise in visitors from Europe. In the last five years cruise ship visitors have increased with a consequent decline in the hotel sector. The government is seeking further investment to develop the tourist trade; such as developing new resort facilities on Paradise Island, in conjunction with a major South African company.

Other revenue is earned from a large registered merchant fleet and from international financial services. The Bahamas is one of the world's largest offshore banking centres.

The government aims to diversify and attract foreign investment. Two industrial sites have been developed, in New Providence and the other in Grand Bahama, where industries include garments, plastic containers, purified water and pharmaceuticals.

THE BAHAMAS	
STATUS	Commonwealth Nation
AREA	13,865 sq km (5,350 sq miles)
POPULATION	269,000
CAPITAL	Nassau
LANGUAGE	English
RELIGION	Anglican Christian majority, Baptist and Roman Catholic minorities
CURRENCY	Bahamian dollar (BSD)
ORGANIZATIONS	Caricom, Comm., OAS, UN

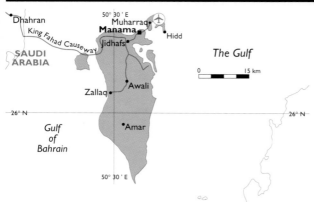

PHYSICAL GEOGRAPHY

The State of Bahrain is an archipelago of 33 islands, all low-lying. The largest is Bahrain Island which consists mainly of sandy plains and salt marshes.

CLIMATE

The summers are hot and humid, especially between April and October when temperatures average 32°C (89°F). The winters are mild. Annual rainfall is less than 76 mm (3 inches).

POLITICS AND RECENT HISTORY

A British protectorate until 1971, Bahrain is now ruled by an Amir from the ruling Al Khalifah family and a National Assembly. Political parties are not permitted.

The state has been disrupted by a series of anti-government demonstrations as Bahrainis have demanded a return to parliamentary rule and some action by the government to relieve the unemployment situation. There is resentment by some Shia muslims who believe the Sunni have preference in job allocation, unemployment being especially high in the majority Shia community.

ECONOMY

Bahrain is an important trading and service centre for the Gulf region. During the Gulf war it supplied the US-led forces stationed in Dhahran. It is also an important banking and financial centre for the Middle East, and a stock exchange, which opened in 1989, extended trading to non-citizens.

Oil was discovered 60 years ago, but at the current rates of extraction the reserves are not expected to last for more than ten years. Oil is still significant in the economy, providing 60 per cent of government revenue. Gas is becoming increasingly important, and reserves are expected to last a further 50 years.

Cheap gas supplies are the basis of Bahrain's aluminium smelting industry, which is the Gulf region's largest non-oil activity. Studies have been undertaken by a Portuguese company for the possible development of an industrial zone and new port at Hidd, and British Gas has been investigating the feasibility of building a power station and desalination plant.

Exports include refined petroleum, chemicals and aluminium. Bahrain's main export partners are the United Arab Emirates, Singapore and Japan.

Tourism is being promoted with the development of hotels and facilities for horse-racing, golf and scuba diving. Approximately 75 per cent of tourists are from neighbouring Arab countries. Tourism and local industries received a significant boost to trade in 1986 when the King Fahad causeway was opened, connecting Bahrain to Saudi Arabia. Hotels benefited with a 40 per cent increase in visitors.

THE STATE OF BAHRAIN	
STATUS	State
AREA	661 sq km (225 sq miles)
POPULATION	539,000
CAPITAL	Manama (Al Manamah)
LANGUAGE	Arabic, English
RELIGION	60% Shi'a and 40% Sunni Muslim, Christian minority
CURRENCY	Bahraini dinar (BHD)
ORGANIZATIONS	Arab League, UN

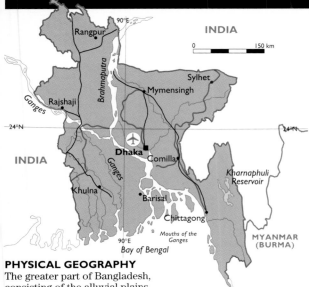

PHYSICAL GEOGRAPHY
The greater part of Bangladesh, consisting of the alluvial plains and deltas of the Ganges and Brahmaputra rivers, is less than 10 m (30 ft) above sea level. The southeast coasts are fringed with mangrove forests.

CLIMATE
The climate is hot, with heavy monsoon rainfall in the summer months.

POLITICS AND RECENT HISTORY
Bangladesh, formerly the Eastern Province of Pakistan, proclaimed its independence in 1971.

Following the resignation of the long-serving President Ershad in 1990, free general elections were held in February 1991. The Bangladesh National Party gained most seats and, supported by a minor party, was able to form a government with its leader Begum Khaleda Zia as prime minister. She immediately proposed a constitutional amendment to establish parliamentary rather than presidential authority. This amendment was approved by parliament in a referendum and was adopted in September 1991. By early 1993 the government, and in particular the prime minister, were criticized for lack of control over law and order and the economy. Since then recriminations between the ruling party and the opposition Awami League, under the leadership of Mrs Sheikh Hasina, have increased. The opposition has boycotted parliament which has thus been paralysed. Attempts by the British Government to mediate have failed.

ECONOMY
The Bangladeshi economy is heavily dependent upon agriculture, and this is subject to the vagaries of climate. Severe cyclones and floods are common during the monsoon season. They occurred in 1988, 1989, 1991 and 1993. The 1993 rains, four times the average and the worst for 40 years, put half the land under water. Despite such disasters agricultural output has been increasing. The chief crop is rice, but wheat is growing in importance. The main cash crop is jute. Bangladesh is the world's leading supplier and it accounts for 25 per cent of the country's exports. Although the value of this commodity has been increasing, long-term prospects are poor because of synthetic competitors.

As one of the world's poorest nations, with an average per capita income under $250 p.a., Bangladesh is heavily dependent on international aid, but there are signs of improvement. The trade deficit is lessening, the International Monetary Fund's fiscal and monetary targets have been met and tax revenues are better than forecast. The government has embarked upon privatization programmes and is encouraging foreign investment although potential investors are discouraged by the political deadlock. Plans exist to emphasise the necessity of birth control, vital if poverty is to lessen.

THE PEOPLE'S REPUBLIC OF BANGLADESH	
STATUS	Republic
AREA	144,000 sq km (55,585 sq miles)
POPULATION	115,203,000
CAPITAL	Dhaka
LANGUAGE	Bengali (Bangla), Bihari, Hindi, English
RELIGION	85% Muslim, Hindu, Buddhist and Christian minorities
CURRENCY	taka (BDT)
ORGANIZATIONS	Col. Plan, Comm., UN

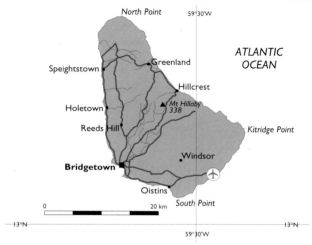

PHYSICAL GEOGRAPHY

Barbados, the most easterly of the Windward Islands, is generally low-lying with some hilly country in the north.

CLIMATE

The climate is pleasant and warm throughout the year; with the temperature ranging from 25–28°C (77–82°F). The heaviest rainfall occurs in the autumn months, with an overall annual rainfall of about 635–1,905 mm (50–75 inches). December to June tend to be dry.

POLITICS AND RECENT HISTORY

Barbados became self-governing in 1961 and fully independent from the UK, as a member of the Commonwealth, in 1966. The head of state is the British monarch, who is represented on the island by a governor-general.

The Barbados Labour Party (BLP) was returned to power in 1994 under the leadership of Owen Arthur. The previous administration of the Democratic Labour Party surrendered power at the election after former prime minister Erskine Sandiford was defeated in a confidence vote.

Barbados is a founder member of Caricom, the Caribbean Community and Common Market, which was established in 1973 to provide a common market in the region, to co-ordinate foreign policy and to co-operate in areas such as education, health, sports and tax administration.

ECONOMY

The Barbadian economy was traditionally founded upon sugar production, but since the 1970s this has been overtaken in importance by tourism. Although sugar is still exported, its value as a proportion of GDP continues to decline and harvests are diminishing. Barbados has one oilfield which supplies 30 per cent of domestic needs, and there are natural gas reserves. The country experienced difficult times in the early 1990's but recent years have seen improved economic performance. Growth has replaced decline and unemployment figures, though still high, have receded.

The government plans to introduce Value Added Tax, a two-year wage freeze is proposed and the World Bank is providing aid. More general incentives have been introduced to encourage much-needed foreign investment.

To improve the tourist trade, which has suffered from world recession, the government is investing in improving facilities. It raises revenue from cruise ships visiting the island by means of a tax on those ships.

Tourism is the main growth sector but other areas of the economy such as data processing and financial services are developing rapidly. Success in these areas is likely because, compared with many other Caribbean territories, the work force is literate and skilled while the infrastructure is well maintained with reliable water, electricity and telecommunications services.

BARBADOS	
STATUS	Commonwealth State
AREA	430 sq km (166 sq miles)
POPULATION	260,000
CAPITAL	Bridgetown
LANGUAGE	English
RELIGION	Anglican Christian majority, Methodist and Catholic minorities
CURRENCY	Barbados dollar (BBD)
ORGANIZATIONS	Caricom, Comm., OAS, UN

PHYSICAL GEOGRAPHY

Belarus is a land of plains and low hills covered with glacial soils and drained mainly by the Dnieper and Dvina river systems. About 30 per cent of the land area is forested; towards the southern border with Ukraine there are extensive marshy tracts lying along the valley of the River Pripyat.

CLIMATE

The climate is continental with warm summers and cold, fairly dry winters.

POLITICS AND RECENT HISTORY

Belarus declared its independence from the Soviet Union in August 1991, following which Stanislav Shushkevich was elected by the Supreme Soviet to be its chairman, albeit with no strong political backing, and held both the position of president as well as that of speaker of the parliament. The issue of neutrality is controversial within Belarus. Shushkevich's views were in conflict with those of the Communist-dominated parliament and Prime Minister V. Kebich. Shushkevich was ousted in early 1994 and replaced temporarily by a nominal head of

state, Mechieslav Grib, although Kebich retained power. However in presidential elections in the summer of 1994 Alexander Lukashenko won a decisive victory over Kebich, winning about 80 per cent of the vote. Lukashenko's platform was closer political and economic ties with Russia. Russia however has shown no great interest in closer ties with its poor neighbour and has specifically ruled out monetary union.

ECONOMY

The Belarussian economy was closely integrated with that of Russia, especially in its dependence on energy supplies. Belarus has almost no oil, gas or coal, although it has substantial peat deposits which are used in power stations and

for domestic consumption. The republic's heavy industries, which include large petrochemical and engineering complexes, were dependent on Russian fuel, and independence has meant economic difficulties. Industrial output has declined and inflation has reached alarming proportions.

Traditionally Belarus has had an agricultural economy based on beef cattle and crops such as hardy grains (rye, oats, buckwheat), sugar-beet, flax and potatoes but even this sector has declined.

In the future Belarus may be forced to concentrate on greater exploitation of the natural wealth of its timber reserves which have traditionally been an important source of economic wealth.

The former government did take some preliminary steps towards a market economy and secured a World Bank loan to assist reform. However little progress has been made nor does any seem likely under the present regime. Severe austerity measures, including steep rises in the price of foodstuffs and old style soviet economies, are the order of the day.

THE REPUBLIC OF BELARUS	
STATUS	Republic
AREA	208,000 sq km (80,290 sq miles)
POPULATION	10,188,000
CAPITAL	Minsk
LANGUAGE	Belorussian, Russian
RELIGION	Roman Catholic, Uniate (Orthodox-rite Roman Catholic)
CURRENCY	zaichik (or zaitsev)
ORGANIZATIONS	CIS, UN

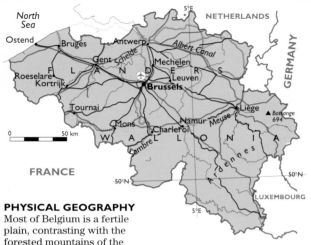

PHYSICAL GEOGRAPHY
Most of Belgium is a fertile plain, contrasting with the forested mountains of the Ardennes in the south.

CLIMATE
The climate is maritime and temperate, with mild winters and cool summers.

POLITICS AND RECENT HISTORY
Since World War II Belgium has established a reputation as a pioneer of international co-operation, as a founder member of the Benelux Union and what is now the European Union (EU) and host to the North Atlantic Treaty Association headquarters. It is the home base for over 800 other international organizations.

Internal affairs have been affected by disunity between the Flemish-speaking north and the French-speaking Walloons to the south. In the 1970s the constitution recognized the two regions of Flanders and Wallonia. However, increasing discontent became evident in 1991 when one million people voted for extremist and anti-political parties in the general election. All the traditional Belgian parties lost seats, and Wilfred Marten's centre-left coalition gave way to another coalition under Jean-Luc Dehaene.

Separatist trends have continued. In the north, increased economic strength combined with a historic sense of identity has further encouraged Flemish nationalism. Meanwhile, the Walloons feel they suffer from minority status, both economically and politically, because the last 11 government leaders have all been Flemish speakers.

Even so these separatist movements were rejected by the Belgian electorate when in May 1995 Jean-Luc Dehaene was again returned to office. This victory gained also in the face of allegations of political corruption confirms an underlying belief by Belgians in their nationhood. This belief is reinforced by the monarchy, repre-

sented now by King Albert who succeeded upon the death of his brother King Baudouin in 1993.

ECONOMY
Belgium is nearly self-sufficient in food production and is heavily industrialized, producing textiles, diamonds, chemicals, machinery and metals. The country is one of the largest car exporters in the world. However the economy has suffered from recession in recent years. There is a serious budget deficit and Mr Dehaene must cut public expenditure despite high levels of unemployment..

There are financial imbalances between Flanders and Wallonia. The Walloon economy, based on coal and steel, has declined. The region is now reckoned to be one of the EU's poorest. The Flemish feel they are subsidizing their poorer partner. However the economy of Flanders remains basically strong and is responsible for 70 per cent of all Belgium's exports.

THE KINGDOM OF BELGIUM	
STATUS	Kingdom
AREA	30,520 sq km (11,780 sq miles)
POPULATION	10,046,000
CAPITAL	Brussels (Bruxelles/Brussel)
LANGUAGE	French, Dutch (Flemish), German
RELIGION	Roman Catholic majority, Protestant and Jewish minorities
CURRENCY	Belgian franc (BEF)
ORGANIZATIONS	Council of Europe, EEA, EU, NATO, OECD, UN, WEU

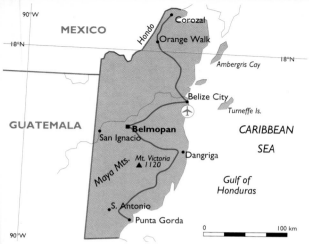

PHYSICAL GEOGRAPHY

Much of the country is jungle, with many rivers and lakes. The Maya mountain range in the southwest is heavily forested. Offshore there are cays and a long coral barrier reef.

CLIMATE

The climate is tropical with a heavy annual rainfall of 1,900 mm (75 inches). Temperatures vary only slightly with seasons, from 23–27°C (73–81°F).

POLITICS AND RECENT HISTORY

When Belize (formerly British Honduras) became independent in 1981, a British garrison was retained as a deterrent to territorial claims by Guatemala. In 1991 Guatemala recognized the independence of Belize, agreed to reduce its maritime zone to three miles and to negotiate for Guatemalan use of Belize's ports. Britain withdrew its troops in 1994.

The Guatemalan president, Jorge Serrano, who had pressed for Belizean recognition, was overthrown in June 1993. After this, and Britain's announcement ending its defence commitment, the Belize government called an election. The opposition, led by Manuel Esquivel, won, probably due to the former prime minister's failure to fight for the retention of the British garrison.

Belize is developing new problems: drug trafficking and other crimes are soaring and the ethnic composition of the country is changing. The Spanish-speaking population, boosted in numbers by an influx of refugees and economic immigrants, was shown in the 1991 census to out-number the native black creoles for the first time. The Spanish speaking population's willingness to work for low rates is creating controversy.

ECONOMY

Belize, famed for its forest products, turned to sugar pro- duction in the 1950s and for the next 30 years became primarily a sugar exporter. In the 1980s, however, world sugar prices fell and the industry declined. In its place came a boom in citrus exports, and now both citrus concentrates and sugar are important exports. Other commodities include bananas, vegetables, tropical fruits, fish and timber. The agricultural sector contributes approximately 75 per cent of Belizean exports and 65 per cent of foreign earnings.

Manufacturing consists primarily of citrus and sugar processing and the preparation of timber from the extensive hardwood forests.

The tourist industry is growing steadily and accounts for 26 per cent of GNP (gross national product). It is an important source of foreign exchange when world prices of cash crops fall. The clearing of coastal mangrove swamps for hotel development has however caused some concern for the environment and protected marine areas have been declared.

BELIZE	
STATUS	Commonwealth Nation
AREA	22,965 sq km (8,865 sq miles)
POPULATION	205,000
CAPITAL	Belmopan
LANGUAGE	English, Spanish, Maya
RELIGION	60% Roman Catholic, 40% Protestant
CURRENCY	Belizean dollar (BZD)
ORGANIZATIONS	CARICOM, Comm., OAS, UN

BENIN

PHYSICAL GEOGRAPHY
Beyond the coastal lowlands the land rises to plateaux culminating in the Atacora Range.

CLIMATE
The climate throughout Benin is hot all year round. Rainfall reaches its maximum in June.

POLITICS AND RECENT HISTORY
Benin (Dahomey until 1975) became independent from France in 1960. Until recently the country was ruled by a one-party system along Marxist-Leninist lines. Constitutional change allowed for multi-party elections in 1991. Nicéphore Soglo defeated the incumbent, ruling with a coalition of ten parties known as *Le Renouveau*. In July 1993 President Soglo approved a new political party, *Renaissance du Benin*, led by his wife.

ECONOMY
Benin is mainly an agricultural economy. Food crops thrived following heavy rains in recent years, but cash crops such as oil palms, cocoa and coffee continue to decline. Cotton growing, though suffering from weak world prices, has expanded. Benin derives some income from limited offshore oil resources. The economy received a boost during the conflict in neighbouring Togo, causing much trade to be transferred from Lomé to Cotonou. There are some light industries but nearly all manufactured goods have to be imported. France is the main trading partner.

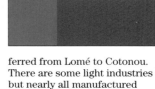

THE REPUBLIC OF BENIN	
STATUS	Republic
AREA	112,620 sq km (43,470 sq miles)
POPULATION	5,215,000
CAPITAL	Porto Novo
LANGUAGE	French, Fon, Adja
RELIGION	majority traditional beliefs, 15% Roman Catholic, 13% Muslim
CURRENCY	CFA franc (W Africa) (XOF)
ORGANIZATIONS	ECOWAS, OAU, UN

BHUTAN

PHYSICAL GEOGRAPHY
Bhutan, in the eastern Himalayas, is mountainous in the north, pre-dominantly forested at its centre and has a tropical lowland to the south.

CLIMATE
The climate varies with altitude but the higher north remains perpetually cold. The southern lowlands are sub-tropical and the centre temperate. Monsoons occur between June and August.

POLITICS AND RECENT HISTORY
The country is effectively an absolute monarchy, responsible for its own affairs, but abides by a treaty with India for guidance on foreign policy. Tradition and the Buddhist culture are preserved and a social code enforces national dress and language. An aggressive stance towards the immigrant population has resulted in a flow of refugees into Nepal and India.

ECONOMY
Bhutan is one of the world's poorest countries and the most rural with only 6 per cent of the population living in towns. Subsistence agriculture with bartering is the rule. Cash crops are fruit and cardamom. Cement, talcum and timber are the main exports. The majority of trade is with India. Stamps and tourism earn foreign exchange but tourism is strictly controlled. India helps to train the army, maintain Bhutan's roads and is involved in the development of hydro-electric power schemes.

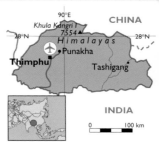

THE KINGDOM OF BHUTAN	
STATUS	Kingdom
AREA	46,620 sq km (17,995 sq miles)
POPULATION	600,000
CAPITAL	Thimphu
LANGUAGE	Dzongkha, Nepali, English
RELIGION	Mahayana Buddhist, 30% Hindu
CURRENCY	ngultrum (BTN), Indian rupee (INR)
ORGANIZATIONS	Col. Plan, UN

PHYSICAL GEOGRAPHY
Most of Bolivia's development is in the high plateau, the Altiplano, within the ranges of the Andes. To the southeast there are semi-arid grasslands and to the north there are dense Amazon forests.

CLIMATE
High western regions experience wide diurnal variations in temperature but without much annual change. The western plains are arid in the south and subject to humid tropical conditions in the north.

POLITICS AND GOVERNMENT
Military rule in Bolivia ended in 1978, since when power has been exercised by civilian governments. In the 1993 presidential elections Gonzalo Sanchez de Lozada of the Revolutionary Nationalist Movement (MNR) overcame the main opposition led by General Hugo Banzer, once a military dictator. Sanchez was installed as president in August 1993.

Despite 15 years of civilian rule, politicians are not well regarded by Bolivians due to previous incompetent administrations, widespread poverty and continuing corruption. A curious feature of Bolivian political life is an obsession with its lack of coastline, lost when Chile annexed territory in 1879. A 6,000 strong navy is maintained in the hope that one day maritime status will be regained.

ECONOMY
Bolivia was once the richest nation in South America, its wealth based on silver and latterly on tin. It is now the poorest. This economic backwardness is primarily due to the sudden collapse of world tin prices in 1985. Tin had been Bolivia's primary export, but the mining industry, though still important, has generally declined and no longer dominates the economy. Agriculture has assumed that position, and the coca bush is the principal crop. External pressure to reduce drug trafficking has led to efforts, supported by substantial aid programmes, to diversify to other crops. These include sugar, coffee, soya and cotton. As a result the importance of coca has declined in recent years although agriculture generally suffers from a lack of mechanisation and a poor communications infrastructure. Despite its poverty recent Bolivian governments have made significant progress and have achieved a measure of stability. In 1985 inflation was 23, 000 per cent but is now comfortably within single figures. Attempts at privatisation were initially half-hearted but are now gaining momentum through a process known as capitalisation, which aims to attract foreign investment into state-owned industry.

Bolivia seeks growth by means of economic links with its neighbours and has preliminary free trade agreements with Brazil, Paraguay, Argentina and Uruguay as a step towards its ambition to become a full member of MERCOSUR. A vital economic development, the construction of a pipeline to supply natural gas to Brazil has been stalled over questions of pricing and respective shares in the project despite agreement in principle by both nations.

THE REPUBLIC OF BOLIVIA	
STATUS	Republic
AREA	1,098,575 sq km (424,050 sq miles)
POPULATION	7,237,000
CAPITAL	La Paz
LANGUAGE	Spanish, Quechua, Aymara
RELIGION	Roman Catholic majority
CURRENCY	Boliviano (BOB)
ORGANIZATIONS	OAS, UN

PHYSICAL GEOGRAPHY

Much of the country is mountainous, with limestone ridges oriented northwest to southeast traversing the country. The only lowlands of consequence are along the valley of the Sava in the north.

CLIMATE

Summer temperatures are warm but in winter, depending upon altitude and aspects, it can be extremely cold.

POLITICS AND RECENT HISTORY

Bosnia-Herzegovina declared its independence from the Federal Republic of Yugoslavia in April 1992 and was recognized by the European Union and the USA shortly thereafter. Almost immediately civil war erupted between the three factions, the Muslims, the Bosnian Serbs and the Bosnian Croats, the latter two elements with support from Serbia and Croatia respectively. In the early months of the civil war the Muslim community suffered territorial losses and the sinister words 'ethnic cleansing' entered the language as the rival factions each outdid the others in barbarous expulsions of large sections of the population from their homes. In June 1992 the United Nations (UN) Security Council voted to deploy UN forces to Bosnia to ensure the supply of humanitarian aid to communities and to protect Sarajevo airport; two months later the authority was extended to allow the use of force to protect relief supplies. In practice the aid supplies were intermittent and subject to protracted delays as the warring factions sought to extract the maximum price and advantage for their cause before allowing the aid to pass.

Numerous efforts to achieve a cease fire failed. The task of the UN forces became ever more futile as they failed to prevent conflict between Bosnian Serbs and Bosnian Muslims. Their parlous position was thrown into sharp focus when Bosnian Serbs took

THE REPUBLIC OF BOSNIA-HERZEGOVINA	
STATUS	Republic
AREA	51,130 sq km (19,736 sq miles)
POPULATION	3,707,000
CAPITAL	Sarajevo
LANGUAGE	Serbo-Croat
RELIGION	Muslim, Christian
CURRENCY	dinar
ORGANIZATIONS	UN

more than 300 UN hostages, following NATO air strikes on ammunition dumps. Following their release, the UN Protection Force withdrew from many vulnerable positions, and in July 1995 the Serbs took the 'safe areas' of Srebrenica and Zepa, amid further widespread accusations of atrocities.

Following a new peace initiative by the U.S. government, the Dayton agreement was signed in December 1995 by the leaders of Bosnia, Croatia and Serbia; Bosnia was to be recognised as a single state with 51 per cent under Muslim/Croat control and 49 per cent Serb. Some territorial adjustments followed in early 1996.

ECONOMY

Before the civil war Bosnia's economy was based predominantly on agriculture, sheep rearing and the cultivation of vines, olives and citrus fruits.

The civil war has ruined the economy. Large sections of the population face desperate hardship and are dependent for survival on UN relief which has itself become ever more sporadic. The currency is worthless; virtually all production has ceased and only the black economy operates.

PHYSICAL GEOGRAPHY

Over half of Botswana consists of the vast upland Kalahari Desert. Elsewhere there are salt-pans and swamps, as in the Okavango Basin.

CLIMATE

The climate is drought-prone with unreliable rainfall of around 538 mm (21 inches) a year. The rain occurs in the hot summer months between October and April. Winters are warm with cold nights and occasional frosts.

POLITICS AND RECENT HISTORY

Botswana, formerly Bechuanaland Protectorate, gained independence from Britain in 1966, since when it has enjoyed a relatively stable democratic government. Politics since independence have been dominated by the Botswana Democratic Party (BDP), which has won every general election with a large majority in the National Assembly. The October 1994 elections were yet again won by the BDP but with a significantly reduced majority, and the incumbent President, Sir Ketumile Masire, was re-elected.

ECONOMY

At the time of independence Botswana was an impoverished state with an economy based on cattle rearing. A year later, in 1967, diamonds were discovered. Wealth from these, and new-found mineral resources, accounted for a massive economic transformation with an average annual growth of 11–13 per cent. Diamonds now account for 50 per cent of GDP and 80 per cent of foreign exchange earnings. The open pit Jwaneng mine is the richest diamond mine in the world in terms of yields. Sales of rough diamonds have risen in recent years and prospects appear good for improved earnings in this sector. Copper-nickel, potash, soda ash, salt and coal are also important. To lessen the dependency on diamonds, manufacturing ventures such as textiles at Selebi-Phikwe are being developed, as is tourism. Visitors to the huge game reserves and national parks are being encouraged by more frequent airline services to Maun in the Okavango, and Maun airport is being improved.

Despite the phenomenal economic growth, there is still much rural poverty and a rapid urbanization trend, resulting in the growth of shanty towns and massive unemployment: 30 per cent of Gaborone's workforce is unemployed.

Botswana's economy is closely linked with neighbouring South Africa, in labour flows, trade (80 per cent of imports are from there) and foreign investment. Future prosperity in Botswana will therefore depend on a relatively stable outcome to South Africa's transition to multiracial democracy.

THE REPUBLIC OF BOTSWANA	
STATUS	Republic
AREA	582,000 sq km (224,652 sq miles)
POPULATION	1,443,000
CAPITAL	Gaborone
LANGUAGE	Setswana, English
RELIGION	majority traditional beliefs, Christian minority
CURRENCY	pula (BWP)
ORGANIZATIONS	Comm., OAU, SADC, UN

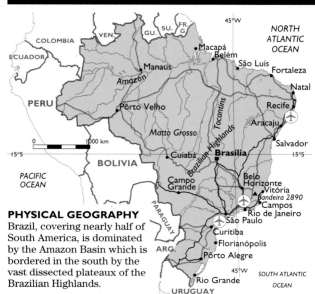

PHYSICAL GEOGRAPHY

Brazil, covering nearly half of South America, is dominated by the Amazon Basin which is bordered in the south by the vast dissected plateaux of the Brazilian Highlands.

CLIMATE

Constant warm and humid conditions in Amazonia give way to seasonal variations towards southern latitudes: winters are cooler with rainfall in summer. The northeast region of Brazil is prone to drought.

POLITICS AND RECENT HISTORY

In 1985 two decades of military rule ended with the formation of a civilian government, and by 1988 the details of the transition to democracy had been drafted in a voluminous new constitution. This included the promise of a plebiscite to be held in 1993 to determine what system of government the country should have: retention of the presidential system, a parliamentary-dominated government or a revival of the monarchy. The result of the referendum, duly held in April 1993, overwhelmingly favoured the first option, endorsing the power of President Itamar Franco who had taken over when Fernando Collor was impeached in late 1992. A revision of the constitution began in October 1993; it is expected to affect the tax system (59 taxes exist), the electoral system (19 parties are represented in Congress) and the government's monopoly of state-owned firms. At the end of 1994 Fernando Henrique Cardoso of the Social Democratic Party succeeded at presidential elections.

ECONOMY

Brazil has a vast variety and wealth of resources, and superlatives abound such as 'the world's leading exporter of iron ore' and 'produces half the world's platinum'. It has one of the fastest growing economies. Agricultural products – notably coffee, sugar, soya beans, and beef – account for 40 per cent of exports. Brazil is also a highly industrialized nation.

Until recently the Brazilian economy had suffered from rampant inflation. President Cardoso, following the introduction of new currency - the Real - has succeeded in a short space of time in conquering the worst of this problem. He is also tackling the unwieldy and inefficient tax system, curbing public spending, reforming and reducing the burden of social security payments and expanding privatisation. The result is a rebirth of the Brazilian economy although there remains a massive disparity between the great wealth of a few and the poverty of many.

Increased trade with Brazil's neighbours will result following the creation of the MERCOSUR customs union and improved exploitation of the nation's natural resources will be possible when planned road networks, controversial from a conservation standpoint, open Amazonia.

THE FEDERATIVE REPUBLIC OF BRAZIL	
STATUS	Federal Republic
AREA	8,511,965 sq km (3,285,620 sq miles)
POPULATION	153,792,000
CAPITAL	Brasilia
LANGUAGE	Portuguese
RELIGION	90% Roman Catholic, Protestant minority
CURRENCY	real
ORGANIZATIONS	Mercosur, OAS, UN

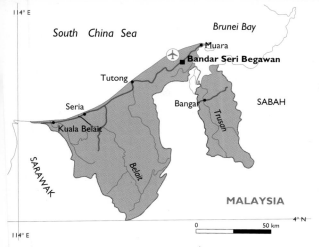

South China Sea

Brunei Bay

Muara

Bandar Seri Begawan

Tutong

Seria

Bangar

SABAH

Kuala Belait

SARAWAK

Belait

Trusan

MALAYSIA

114° E

114° E

4° N

0   50 km

PHYSICAL GEOGRAPHY
Brunei is a small forested state, mostly lowland but with mountains that reach 1,800 m (5,900 ft) in the southeast.

CLIMATE
The climate of Brunei is tropical and humid with temperatures all year round in the range 26–28°C (79–82°F). There is heavy rainfall of 2,500–5,000 mm (100–200 inches) a year, especially inland.

POLITICS AND RECENT HISTORY
Brunei, a former British protectorate, became an independent sovereign state in 1984 and is now an absolute monarchy, ruled by its Sultan who is said to be the world's richest man. Political parties are banned and attempts to form an opposition have been denied. The Sultan appoints a council of ministers to assist his government, which seeks to emphasize its Islamic credentials. The recent growth of Islamic influence in the state has led to emigration of Chinese members of the community. In recent years signs of some dissent have been expressed in the form of anonymous circulars, distributed in the university and

offices in the capital. These criticized the government for incompetence and the growing apparent wealth of government members. Another reason for discontent was that there had been no pay rise in the civil service for eight years, whereas wealth was obvious in the building of new palaces in royal circles.

Brunei's relationship with its neighbour, Malaysia, has sometimes been uneasy. The declaration by Brunei of a 200-mile exclusive economic zone brought a Malaysian response that the Brunei border would transgress its own exclusive zone. A border dispute concerning the Limbang district of Sarawak, claimed by Brunei, is holding up the development of the Pan-Borneo highway, scheduled to link Sarawak,

Brunei and Sabah. However, Malaysia and Brunei have recently agreed to set up a Border Committee, and better relations are encouraged with joint land military exercises.

ECONOMY
Brunei's economy is founded almost entirely on its oil and gas reserves. The oil and gas wealth has provided the small population with living standards almost without rival. There is no income tax, and a comprehensive welfare system is available to all. Continued prosperity is expected. Brunei's major trading partner is Japan, which receives 90 per cent of all liquefied natural gas sold. A new 20-year contract to supply 5.54 million tons of liquefied natural gas annually has recently been signed between Brunei and Japan.

The Government is making efforts to diversify the economy by offering tax concessions to foreign investors to encourage industries such as petrochemicals, glass, timber, paper and fertilisers.

BRUNEI DARUSSALAM	
STATUS	Sultanate
AREA	5,765 sq km (2,225 sq miles)
POPULATION	274,000
CAPITAL	Bandar Seri Begawan
LANGUAGE	Malay, English, Chinese
RELIGION	65% Sunni Muslim, Buddhist and Christian minorities
CURRENCY	Brunei dollar (BND)
ORGANIZATIONS	ASEAN, Comm., UN

PHYSICAL GEOGRAPHY

The Rhodope mountains dominate the west while the Balkan mountains (Stara Planina) form a chain running east–west through central Bulgaria. To the north lie the plains of the Danube, and to the south are the lowlands of Thrace and the Maritsa valley.

CLIMATE

Temperatures show significant seasonal variation, with summer averages exceeding 20°C (68°F) and winter averages around 0°C (32°F). Rainfall maxima are in summer.

POLITICS AND RECENT HISTORY

Following the collapse of the Communist regime of President Zhivkov, who was ousted in 1989, and briefly, Petar Mladenov in 1990, free elections resulted in success for the Union of Democratic Forces (UDF), an anti-Communist coalition. Their leader, Zhelyu Zhelev, was victorious in presidential elections in early 1992 following the formation of a government under prime minister Lyuben Berov in October 1991. Zhelev, although under pressure from the UDF has survived in power but Berov, following the loss of several confidence votes, resigned in September 1994. In general elections three months later the Socialist Party, mainly comprising former communists, won an overall majority under leader Mr Zhan Videnov. Even so a substantial proportion of Bulgarians favour a restoration of the monarchy as a catalyst for stability and it is thus possible that King Simeon, living in exile in Spain, may take the throne.

ECONOMY

The basis of the Bulgarian economy is agriculture. Fertile lands in theory allow the cultivation of a wide variety of crops including cereals, vines, cotton, tobacco and fruit. However, the agricultural sector has not made a successful transition from Communist rule. The state farms passed into private ownership but mostly as very small plots. The owners, unable to command the resources necessary for the investment in machinery and fertilizers that could achieve yields comparable to those of western Europe, have resorted in many cases to cooperative methods.

Industrial output has also declined since the downfall of communism but is showing signs of recovery. A significant factor in these difficulties is Bulgaria's adherence to the United Nations sanctions on trade with Serbia and Montenegro, which are estimated to have cost Bulgaria millions of dollars. Economic reform in Bulgaria has been conducted at a slow pace although privatisation by means of a voucher system began in 1994. The country's relationship with the IMF (International Monetary Fund) is critical to its development and loans are subject to strict conditions. Bulgaria has agreed to join EFTA (European Free Trade Association) although the effect may be insignificant. It maintains an economic relationship with Russia and has reached an agreement over construction of a pipeline from the Black Sea port of Burgas to the Greek port of Alexandroupolis. This will take oil from the Russian pipeline terminal at Novorossiysk.

THE REPUBLIC OF BULGARIA	
STATUS	Republic
AREA	110,910 sq km (42,810 sq miles)
POPULATION	8,452,000
CAPITAL	Sofia (Sofiya)
LANGUAGE	Bulgarian, Turkish
RELIGION	majority Eastern Orthodox, Muslim
CURRENCY	lev (BGL)
ORGANIZATIONS	Council of Europe, OSCE, UN

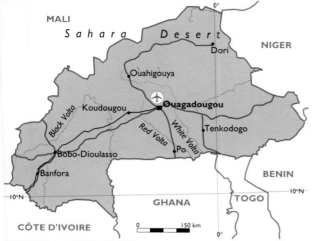

PHYSICAL GEOGRAPHY

The north of the country lies in the Sahara and is thus very arid. The south is largely savannah. Three large rivers, the Black, White and Red Voltas, flow through the country.

CLIMATE

Burkina has a tropical climate. The months from November to February experience dry, cool weather. Rainfall is erratic and the region is prone to severe droughts. The rain months are March and April, with most rain in the southeast.

POLITICS AND RECENT HISTORY

Previously known as Upper Volta, Burkina achieved full independence from France in 1960. Since then there have been a series of military coups. In 1984 Captain Thomas Sankara seized power and introduced a revolutionary nationalist movement, renaming the country Burkina Faso and initiating literacy programmes and afforestation schemes. Internal disputes within the ruling military council, however, led to a confrontation with Sankara's close friend, Blaise Compaore, and on 15 October 1987 Sankara was assassinated. Compaore

succeeded as president and in June 1991 introduced a new constitution allowing many political parties. In the following presidential and legislative elections, however, the opposition parties boycotted the elections, rendering Compaore's victory insignificant. Nevertheless, the multi-party system appeared to meet with some French approval in that President Compaore has received an invitation for his first official visit to France.

The prime minister of Burkina, Youssouf Ouedraogo has also visited Brussels to persuade the European Union (EU) that his country, now ruled by a democratic government, was eligible for EU aid.

ECONOMY

Agriculture, despite its paucity, remains a major source of

income and, together with livestock herding, involves 90 per cent of the population. River blindness, affecting the populations around the Black Volta tributaries, has been eradicated, releasing fertile land for cultivation and attracting farmers from the arid north.

Exports include cotton, groundnuts and some minerals. Manganese exports began in May 1993 and zinc production is scheduled to begin at Perkoa. However gold mining has expanded rapidly in recent times and, even though much is smuggled, it is now the most important export. Even so imports are double the value of exports, reflecting the perennial problem that Burkina has not achieved self-sufficiency in food. Since agricultural production varies with the amount and pattern of annual rainfall, output is extremely unpredictable. Foreign aid is necessary, and an important source of revenue is the remittances from Burkina's migrant workers in neighbouring Côte d'Ivoire and Ghana. Progress is hampered by poor adult literacy which, at around 20 per cent, is among the lowest in the world. However significant investment in schools will in time effect an improvement.

THE REPUBLIC OF BURKINA	
STATUS	Republic
AREA	274,122 sq km (105,811 sq miles)
POPULATION	9,889,000
CAPITAL	Ouagadougou
LANGUAGE	French, More (Mossi), Dyula
RELIGION	60% animist, 30% Muslim, 10% Roman Catholic
CURRENCY	CFA franc (W Africa) (XOF)
ORGANIZATIONS	ECOWAS, OAU, UN

BURUNDI

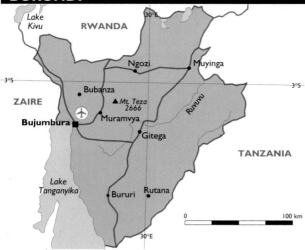

PHYSICAL GEOGRAPHY
Burundi is hilly, with high plateaux in the centre and savannah in the east.

CLIMATE
Temperatures, hot near Lake Tanganyika, are cooler elsewhere. Rainfall occurs between October and May.

POLITICS AND RECENT HISTORY
Independent from Belgian administration since 1962, Burundi (formerly part of Ruanda-Urundi) is the second most densely populated country in Africa.

The country has a history of conflict between the Tutsi tribes, who are originally pastoralist and make up 14 per cent of the population, and the traditionally agriculturist Hutu, 83 per cent of the population. For 30 years the country had been Tutsi-governed. Burundi has been notorious for its massacres, the worst occurring in 1972 when 100,000 Hutu were killed as a reprisal for an attempted coup.

In June 1993, in Burundi's first multi-party election, Hutu Melchior Ndadaye won a landslide victory, bringing an end to Tutsi political domination.

Tutsi members were nevertheless represented in government and the army remained under the control of the Tutsi tribe.

However, in October 1993 President Ndadaye became the victim of Burundi's fourth coup since independence. He was overthrown and executed by Tutsi paratroopers. Four senior colleagues were also murdered, and the surviving government members retreated into the shelter of foreign embassies. Violent and widespread slaughter resulted as Tutsis attacked Hutus and Hutus took revenge for the murder of the president. In April 1994 Burundi's new president and the Rwandan president were killed when their plane was shot down as they returned from discussions held in Tanzania, aimed at ending the ethnic conflict. This sig-

nalled the outbreak of violence in Rwanda and continuing conflict in Burundi. Sylvestre Ntibantunganya, a Hutu, was appointed president in late 1994 but commands little authority and has been accused of making too many concessions to the Tutsis. United Nations efforts at mediation and the creation of a government of National Unity formed from the Front for Democracy in Burundi (FRODEBU) Hutu party and the Union for National Progress (UPRONA) Tutsi party have not succeeded. The prime minister has resigned and massacres continue.

ECONOMY
Coffee remains Burundi's chief export, although tea and cotton are also produced. There has been a significant growth in manufacturing output but now tribal strife has ruined the economy as the population moves en masse from the towns and villages. Burundi always was one of the world's poorest countries and it is now even poorer. It will for many years rely on the international community and foreign aid for survival.

THE REPUBLIC OF BURUNDI	
STATUS	Republic
AREA	27,835 sq km (10,745 sq miles)
POPULATION	5,958,000
CAPITAL	Bujumbura
LANGUAGE	French, Kirundi, Swahili
RELIGION	60% Roman Catholic, animist minority
CURRENCY	Burundi franc (BIF)
ORGANIZATIONS	OAU, UN

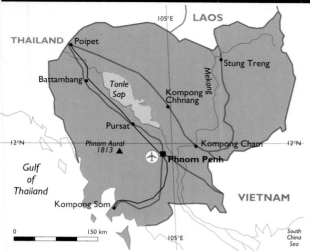

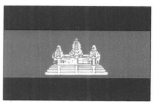

PHYSICAL GEOGRAPHY
Most of the country is a low-lying land basin drained by the Mekong river, with the Tonle Sap (Great Lake) at its heart.

CLIMATE
The climate is tropical, with average temperatures exceeding 25°C (77°F) throughout the year. Monsoon rainfall occurs from May to October.

POLITICS AND RECENT HISTORY
Cambodia, known as the Khmer Republic from 1970 to 1976 and Kampuchea until 1991, has suffered civil war and foreign invasion since 1970 when Prince Sihanouk was overthrown by a right wing coup. Five years of civil war followed until 1975, when the Maoist Khmer Rouge gained power under Pol Pot and embarked upon a brutal episode of genocide. Vietnamese troops invaded in 1978 and installed a left wing government, but the Khmer Rouge continued to exert powerful influence as a ruthless guerrilla movement. Vietnamese withdrawal in 1989 was followed by further fighting until the United Nations (UN), supported by China and

Russia, achieved a peace agreement – the Paris Accord – in October 1991. According to this agreement, a UN Transitional Authority in Cambodia was to supervise an interim government and oversee democratic elections.

The elections, boycotted by the Khmer Rouge amid threats of violence, were held in May 1993. The victor was the royalist Funcinpec party, closely followed by the left wing People's Revolutionary Party (PRP). After some wrangling, the two parties formed a coalition with Prince Ranariddh of Funcinpec as prime minister and Hun Sen of the PRP as deputy. The administration drafted a new constitution, creating a democratic monarchy. Although it was criticized by observers, particularly for its lack of

emphasis on human rights, the new constitution was approved by congress, and in October 1993 Prince Sihanouk was reinstated as king.

Since the elections, Khmer Rouge power has waned, although it is still active. The government has retained power but has lost the services of its able foreign and finance ministers, who refused to support the outlawing of the Khmer Rouge, believing it a futile gesture. Moreover, there are widespread allegations of corruption within the government.

ECONOMY
Years of strife left Cambodia bankrupt, without budget or payment of taxes. Until his dismissal, Sam Rainsy, the finance minister, had begun to improve the economy. He reduced inflation, improved revenue collection, cut public spending and stabilized the exchange rate.

Reconstruction will take many years – the country's infrastructure must be rebuilt and agricultural land brought back into production. However, Cambodia is receiving foreign investment, mostly from Malaysia.

THE KINGDOM OF CAMBODIA	
STATUS	Kingdom
AREA	181,000 sq km (69,865 sq miles)
POPULATION	9,308,000
CAPITAL	Phnom Penh
LANGUAGE	Khmer
RELIGION	Buddhist majority, Roman Catholic and Muslim minorities
CURRENCY	Riel (KHR)
ORGANIZATIONS	Col. Plan, UN

CAMEROON

PHYSICAL GEOGRAPHY

Along the border with Nigeria there are volcanic mountains, the highest being Mt. Cameroun at 4,095 m (13,435 ft). The northern part of the country reaches into desert; further south there is savannah, and then tropical rainforest. The coastal strip is cultivated.

CLIMATE

The climate is equatorial, with plentiful rainfall in the range 1,570–4,050 mm (62–160 inches), generally most abundant between July and October and declining inland. The south is hot and dry between November and February.

POLITICS AND RECENT HISTORY

Cameroon has been independent since 1961, and has been ruled as a one-party state for most of that time. The first multi-party elections were held in October 1992 and, when the government retained control, there was an outcry of election-rigging. The head of state is President Paul Biya, who has been in power since 1982. During 1994, an alliance of opposition groups was established under the leadership of John Fru Ndi, who was replaced by Samuel Eboua in 1995.

An increasing cause of contention in Cameroon is the rivalry between the minority (20 per cent) English speakers and the majority French speakers. Both languages are official, but the English speakers are clamouring for autonomy, claiming French-speaking domination of the country in the public sector. President Biya drafted a new constitution, but this rules out any return to the country's pre-1972 federalist structure.

In August 1993 a meeting was held in Yaoundé of the Cameroon–Nigeria Joint Border Commission. The Commission discussed the demarcation of the respective borders and the development of resources within the border zones. However, in 1994 a dispute between Cameroon and Nigeria developed over the ownership of the oil-rich peninsula of Bakassi.

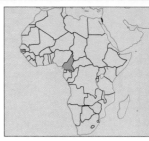

ECONOMY

Despite Cameroon's political battles, the country achieves self-sufficiency in food. Agriculture, based on plantations, is strong with a variety of produce – mainly cocoa, coffee, palm and rubber. In recent years, bananas have become a leading cash crop and avocados an important export commodity. France and the Netherlands are the major export markets.

Oil products, once the main exports, continue to decline. Cameroon has considerable bauxite deposits. Aluminium is smelted at Edea, in a major smelting complex, using hydroelectric power. Further deposits of bauxite and cassiterite occur in the Adamaoua plateau, but these are yet to be exploited.

The economy, however, is shrinking and the income per head has halved since 1985. An International Monetary Fund programme lapsed in 1992.

THE REPUBLIC OF CAMEROON	
STATUS	Republic
AREA	475,500 sq km (183,545 sq miles)
POPULATION	12,522,000
CAPITAL	Yaoundé
LANGUAGE	English, French
RELIGION	40% Christian, 39% traditional beliefs, 21% Muslim
CURRENCY	CFA franc (C Africa) (XAF)
ORGANIZATIONS	Comm, OAU, UN

Map labels: Lake Chad, 300 km, Maroua, 10°E, 10°N, NIGERIA, Garoua, CHAD, Adamaoua Plateau, Ngaoundéré, Foumban, Mamfé, Sanaga, CENTRAL AFRICAN REPUBLIC, Kumba, Nkongsamba, Bakassi Pen., Mt. Cameroun 4095, Buéa, Douala, Edea, Bertoua, Yaoundé, Gulf of Guinea, Ebolowa, EQ.GUINEA, 10°E, GABON, CONGO

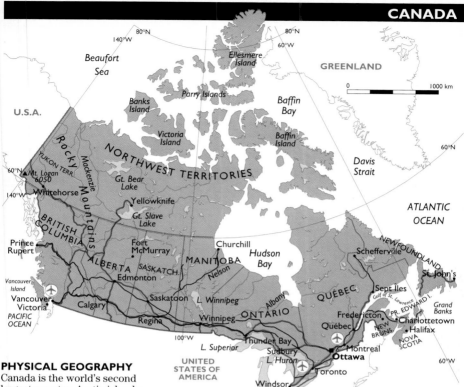

PHYSICAL GEOGRAPHY

Canada is the world's second largest country. Arctic islands and tundra give way southwards to boreal forests, interspersed with lakes and mighty rivers, and then central prairies. To the east are the fertile lowlands of the St. Lawrence basin. The Rocky mountains rise to over 4,000 m (13,000 ft) in the west, beyond which are the coastal mountains, fjords and islands of British Columbia

CLIMATE

Generally the climate is continental, with severe winters and hot summers, but considerable variations occur with latitude, altitude and maritime influence.

POLITICS AND RECENT HISTORY

Traditionally the federal government of Canada has been controlled by either the Liberals or (during the last nine years) the Progressive Conservatives, with the New Democrats a strong third. However, in the general election of October 1993 the Progressive Conservative party's holding of 157 seats was reduced to two, the New Democrats' 44 seats to eight, and Jean Chrétien of the Liberals became prime minister in a landslide victory. Two parties rose to new prominence. The Reform Party and the Bloc Québécois gained 52 and 54

CANADA	
STATUS	Commonwealth Nation
AREA	9,922,385 sq km (3,830,840 sq miles)
POPULATION	29,248,000
CAPITAL	Ottawa
LANGUAGE	English, French
RELIGION	46% Roman Catholic, Protestant and Jewish minorities
CURRENCY	Canadian dollar (CAD)
ORGANIZATIONS	Col. Plan, Comm., G7, OAS, OECD, NATO, NAFTA, UN

seats respectively. Jean Chrétien and the Liberals are dedicated federalists and will have to contend with the French-speaking Bloc Québécois, now the official opposition, which aspires for sovereignty for Québec. This cause was reinforced by the victory of their ally, the Parti Québécois, in provincial elections in Québec. After a disagreement about timing, a referendum was held in October 1995 on whether Quebec should secede from Canada. The narrowness of the vote against, however, led to further speculation that the issue is not finally settled.

Meanwhile the Canadian government has reached agreement with the native Inuit population over their land claims in the Northwest Territories. On 1 April 1999 the territory of Nunavut, comprising one-fifth defined as belonging to the Inuit.

ECONOMY

Canada has one of the highest standards of living in the world, partly due to the huge mineral reserves. It is a leading producer of zinc, uranium, nickel, copper and other minerals. Exploration has been restricted recently due to environmental constraints, indigenous people's land claims and new taxes. However, valuable diamond deposits, which will be exploited from 1997, have been discovered in the Northwest Territories and massive reserves of nickel, cobalt and copper have recently been located at Voisey Bay, Labrador.

Canada's prairies provide 20 per cent of the world's wheat, and the forests feed major lumber, pulp and paper industries. The Newfoundland fishing industry has been stricken by vanishing stocks in the Grand Banks, thought to be caused by overfishing. This resulted in conflict with the European Union – and Spain in particular – over stocks, especially of Greenland halibut. The conflict was resolved after prolonged negotiations over fishing quotas.

The country's vast rivers provide huge amounts of hydro-electric power but most industry is confined to the Great Lakes and St. Lawrence margins.

The economy has suffered two years of severe recession, although some recovery is evident and inflation is below 2 per cent. A 3 per cent growth rate is forecast.

Canada is a great trading and industrialized nation and has received a boost to trade with the implementation of the North American Free Trade Agreement. Although, in common with other developed countries, Canada has suffered from world recession, the economy has continued to grow and its future wealth is assured.

ALBERTA	
STATUS	**Province**
AREA	**661,190 sq km**
	(255,220 sq miles)
POPULATION	**2,577,000**
CAPITAL	**Edmonton**

BRITISH COLOMBIA	
STATUS	**Province**
AREA	**948,595 sq km**
	(366,160 sq miles)
POPULATION	**3,342,000**
CAPITAL	**Victoria**

MANITOBA	
STATUS	**Province**
AREA	**650,090 sq km**
	(250,935 sq miles)
POPULATION	**1,098,000**
CAPITAL	**Winnipeg**

NEW BRUNSWICK	
STATUS	**Province**
AREA	**73,435 sq km**
	(28,345 sq miles)
POPULATION	**728,000**
CAPITAL	**Fredericton**

NEWFOUNDLAND	
STATUS	**Province**
AREA	**404,520 sq km**
	(156,145 sq miles)
POPULATION	**576,000**
CAPITAL	**St John's**

NORTHWEST TERRITORIES	
STATUS	**Territory**
AREA	**3,379,685 sq km**
	(1,304,560 sq miles)
POPULATION	**56,000**
CAPITAL	**Yellowknife**

NOVA SCOTIA	
STATUS	**Province**
AREA	**55,490 sq km**
	(21,420 sq miles)
POPULATION	**910,000**
CAPITAL	**Halifax**

ONTARIO	
STATUS	**Province**
AREA	**1,068,630 sq km**
	(412,490 sq miles)
POPULATION	**10,168,000**
CAPITAL	**Toronto**

PRINCE EDWARD ISLAND	
STATUS	**Province**
AREA	**5,655 sq km**
	(2,185 sq miles)
POPULATION	**131,000**
CAPITAL	**Charlottetown**

QUÉBEC	
STATUS	**Province**
AREA	**1,540,680 sq km**
	(594,705 sq miles)
POPULATION	**6,954,000**
CAPITAL	**Québec**

SASKATCHEWAN	
STATUS	**Province**
AREA	**651,900 sq km**
	(251,635 sq miles)
POPULATION	**993,000**
CAPITAL	**Regina**

YUKON TERRITORY	
STATUS	**Territory**
AREA	**482,515 sq km**
	(186,250 sq miles)
POPULATION	**29,000**
CAPITAL	**Whitehorse**

PHYSICAL GEOGRAPHY

The country is an archipelago of ten main islands and several smaller ones. They lie some 600 km (375 miles) from Africa and are of volcanic origin. The highest point, Mt. Fogo, approaches 3,000 m (10,000 ft) and is an active volcano.

CLIMATE

Despite its maritime position the islands are arid. Such rain that does occur falls in August and September, although the highest parts receive some at other times of the year. Temperatures are warm throughout the year and seldom fall much below 20°C (68°F) at sea-level.

POLITICS AND RECENT HISTORY

The nation achieved its independence from Portugal in 1975. In its campaign for independence it was associated with Guinea-Bissau, but the constitution adopted in 1981 deleted all reference to possible union. At that time the socialist Partido Africano da Independencia de Cabo Verde (PAICV) was the only legitimate party, but in 1990 the National Assembly overturned one-party rule. At multi-party elections held in the following January, the Movement for Democracy won a decisive victory over the PAICV and a month later their candidate was elected to the presidency.

ECONOMY

Cape Verde has a weak agricultural economy. Output is adversely affected by poor soils, drought and lack of surface water, to the extent that over 90 per cent of foodstuffs must be imported, mostly from Portugal and the Netherlands. Only in the higher regions are conditions adequate to support the growth of some maize, sugar cane, coffee and fruit, mostly bananas. The island of Fogo is the centre of fruit production.

The principal source of export revenue is from fishing, especially tuna, and fish processing and canning have been the only significant industries on the islands. Harsh conditions have driven many of the population abroad, and remittances from expatriates are the single most important contribution to sustaining the economy.

Some recent economic progress is evident. The first international trade fair was held in Praia in June 1993 and demonstrated the government's intention to encourage investment, especially from Portugal. The fair did indeed attract Portuguese interest. Projects which have been initiated include a gravel extraction enterprise, a wine-bottling factory and a hotel, which may signal the development of a hitherto modest tourist industry.

THE REPUBLIC OF CAPE VERDE	
STATUS	**Republic**
AREA	**4,035 sq km (1,560 sq miles)**
POPULATION	**370,000**
CAPITAL	**Praia**
LANGUAGE	**Portuguese, Creole**
RELIGION	**98% Roman Catholic**
CURRENCY	**Cape Verde escudo (CVE)**
ORGANIZATIONS	**ECOWAS, OAU, UN**

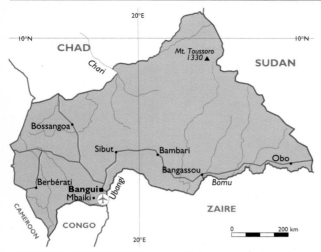

PHYSICAL GEOGRAPHY
Most of the Central African Republic is plateaux covered by scrub or savannah. These are drained in the south by the Ubangi river system and in the north by the Chari, which flows into Lake Chad.

CLIMATE
The climate is hot all the year round with moderately heavy rainfall for much of the year, between 1,417–1,525 mm (57–61 inches). There is a pronounced dry season from December to February.

POLITICS AND RECENT HISTORY
The Central African Republic, formerly part of French Equatorial Africa, gained its independence in 1960. Its early years were notable for the idiosyncratic and corrupt rule of the self-styled Emperor Bokassa. He was deposed in 1979 and subsequently imprisoned following commutation of a death sentence. Since Bokassa's downfall, the country has been subject to one-party rule by the *Rassemblement Démocratique Centrafricaine*, with André Kolingba as president from 1981.

Under pressure from France, the government took slow steps towards political liberalization. Multi-party elections for the presidency were held in October 1992 but were annulled because of irregularities. In the presidential elections eventually held in September 1993, Ange-Félix Patasse won, heading the Central African People's Liberation Party. Following a referendum, a new constitution was adopted in 1995, whereby a president may be re-elected, and not restricted to one six-year term of office. The new constitution also states that the prime minister will implement policies decided by the president. This has caused the opposition some concern, as they can see the balance of power shifting further to favour the

president.

ECONOMY
The economy is dominated by subsistence agriculture, although some cotton, coffee and groundnuts are produced for export. Recent increases in cotton and coffee prices have been an incentive to farmers to cultivate the land. Diamonds and small quantities of gold are the principal source of wealth and account for more than half the foreign earnings. Hardwood forests in the southwest provide timber, also for export.

The Central African Republic is an extremely poor country. The economy has declined since independence and the budget deficit has grown massively. However, credit from the International Monetary Fund was approved in 1994 to support economic reforms.

THE CENTRAL AFRICAN REPUBLIC	
STATUS	**Republic**
AREA	**624,975 sq km (241,240 sq miles)**
POPULATION	**3,173,000**
CAPITAL	**Bangui**
LANGUAGE	**French, Sango (national)**
RELIGION	**animist majority, 33% Christian, Muslim minority**
CURRENCY	**CFA franc (C Africa) (XAF)**
ORGANIZATIONS	**OAU, UN**

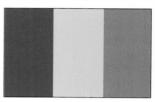

PHYSICAL GEOGRAPHY

Much of the country is plateaux sloping westwards to Lake Chad, the area of which is subject to seasonal fluctuations but is progressively drying up. The highest points, at over 3,300 m (10,800 ft), are in the volcanic Tibesti mountains in the north.

CLIMATE

Climatic conditions range from desert in the north, at the southern edge of the Sahara, through savannah to tropical forest in the southwest. Rainfall in the south averages 990 mm (39 inches) a year. Temperatures range between 24–28°C (75–82°F).

POLITICS AND RECENT HISTORY

Chad has been subject to military or one-party rule since its independence from France in 1960. The government of Hissène Habré was overthrown by the Popular Salvation Movement in November 1990 and its leader, Idriss Déby, declared himself president. Since then some evidence of political transformation has emerged. Political parties were authorized in 1992 and in early

1993 a Sovereign National Conference generated a Transition Charter. Although this retained Idriss Déby as head of state and the army, it proposed a new constitution within 12 months, allowing for multi-party elections and civilian rule. In March 1995, the transitional period was extended for a further 12 months. In preparation for the elections, Chad is preparing a census of voters.

The prospects for political stability are uncertain as strife continues between religious factions – Muslims from the north versus Christians in the south, and nomadic herders versus sedentary crop farmers.

However, on the international front, a dispute between Chad and Libya over the

THE REPUBLIC OF CHAD	
STATUS	Republic
AREA	1,284,000 sq km (495,625 sq miles)
POPULATION	6,098,000
CAPITAL	Ndjamena
LANGUAGE	French, Arabic, local languages
RELIGION	50% Muslim, 45% animist, 5% Christian
CURRENCY	CFA franc (C Africa) (XAF)
ORGANIZATIONS	OAU, UN

Aouzou Strip has been resolved. The 42,000 square mile region, which is thought to contain oil and uranium, had been occupied by Libya since 1973 and twice been the reason for fighting between the two countries. In February 1994 the International Court of Justice ruled it belonged to Chad and several months later Libya completed its withdrawal.

ECONOMY

Chad is one of the world's poorest nations, with a heavy dependency on subsistence agriculture and cattle herding. International food aid is required.

The most important cash crop is cotton, grown in the south and southwest. This provides 50 per cent of Chad's foreign exchange earnings, but output has been adversely affected by civil unrest and drought. Fishing in Lake Chad has declined as the waters have receded.

There are few other natural resources of consequence. Natron is extracted north of Lake Chad, and some oil is produced, although production has remained minimal due to the political unrest.

PHYSICAL GEOGRAPHY

Chile's eastern border is the High Andes, whose peaks exceed 6,000 m (19,500 ft). A coastal range fringes most of the Pacific coast.

CLIMATE

The extreme latitudinal extent, combined with the range of altitude east to west, provides dramatic climatic variation. The north is arid, central parts have a mild Mediterranean climate and the southern extremes are cold, wet and stormy.

POLITICS AND RECENT HISTORY

For 17 years, following the murder of President Salvador Allende in 1973, Chile was governed by the military dictatorship of General Augusto Pinochet. In March 1990 civilian rule returned under Patricio Aylwin, a Christian Democrat, leading a centre-left coalition. The return to civilian rule was achieved on condition that the military retained an influence on government and remained independent of the civilian authorities. General Pinochet holds his post as Army chief and each of the services appoints one of a group of senators. This constitutional position remains under the presidency of Eduardo Frei, who replaced President Aylwin following a landslide victory in 1993, but expires in 1997, when constitutional reform limiting the authority of the armed forces is likely.

ECONOMY

For all the criticisms of Pinochet's rule, he did introduce a capitalist revolution, establishing a measure of prosperity that has continued under civilian rule. President Aylwin used the wealth created to fund social programmes and lift much of the population out of poverty.

The mainstay of the Chilean economy is primary products, especially mineral ores and timber products. Manufacturing is well developed, fruit farming is on the increase and a relatively new salmon industry is earning revenue.

Chile has one of the soundest economies in Latin America – with steady growth and falling inflation – and its stability has attracted significant foreign investment.

Chile is likely to join MERCOSUR and there are tentative plans for major road construction, from the Pacific to the Atlantic, between Chile and Argentina, despite lingering mutual suspicion. Chile will also start discussions on membership of the North American Free Trade Association.

CHILE	
STATUS	Republic
AREA	751,625 sq km (290,125 sq miles)
POPULATION	14,026,000
CAPITAL	Santiago
LANGUAGE	Spanish
RELIGION	85% Roman Catholic, Protestant minority
CURRENCY	Chilean peso (CLP)
ORGANIZATIONS	OAS, UN

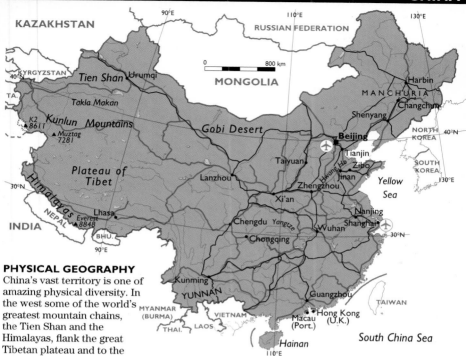

PHYSICAL GEOGRAPHY

China's vast territory is one of amazing physical diversity. In the west some of the world's greatest mountain chains, the Tien Shan and the Himalayas, flank the great Tibetan plateau and to the north the arid Takla Makan basin. Northeast and eastern China is a vast plain of fertile wind-blown glacial soils, while to the south lie the forested mountains of Yunnan.

CLIMATE

Climatic conditions of all kinds are found within China's borders. Extreme continental conditions occur in northern China, where summers are hot with some rainfall and winters are bitingly cold and dry. By contrast the southwest enjoys a moist, warm subtropical climate. Between these two extremes much of the nation experiences temperate conditions.

POLITICS AND RECENT HISTORY

China is a Communist one-party state in which power is formally held by a Central Committee, elected by the Party Congress. The Committee in turn elects a seven-person Politburo. However, the supreme authority is the ailing Deng Xiaoping who came to power in 1976 following the death of Mao Zedong. Deng's authority is such that he does not hold any office within the government. His eventual successor is most likely to be Jiang Zemin, the President and General Secretary of the Communist Party. Other candidates are Li Peng, the Prime Minister, Zhu Rongji, Deputy Prime Minister with responsibility for the economy, and

THE PEOPLE'S REPUBLIC OF CHINA	
STATUS	People's Republic
AREA	9,597,000 sq km (3,704,440 sq miles)
POPULATION	1,200,000,000
CAPITAL	Beijing (Peking)
LANGUAGE	Mandarin Chinese, regional languages
RELIGION	Confucian, Buddhist, Taoist, Christian and Muslim minorities
CURRENCY	yuan (CNY)
ORGANIZATIONS	UN

Qiao Shi, Chairman of the Chinese People's Congress.

China is emerging from relative isolation and has gradually improved its political relationships with other nations. The Tiananmen Square massacre in 1989 has proved a significant stumbling block, as has China's continued occupation of Tibet and the repression of the Tibetan people. China's human rights record has been the subject of widespread criticism, although the government has resolutely maintained that its internal affairs are its own business. Despite this, China has been able to maintain its 'most favoured nation' status with the United States. Relations with the West are nevertheless somewhat fragile because of China's willingness to sell its missiles and nuclear technology to countries such as Iran. Moreover, while declaring its support for the nuclear non-proliferation treaty and an

eventual nuclear test ban, China continues with underground nuclear tests.

In 1997 the British colony of Hong Kong will revert to Chinese sovereignty. Negotiations between Britain and China, designed to achieve a smooth transition, have proved difficult. The main area of controversy has been Britain's attempt to widen democracy through a Legislative Council.

ECONOMY

Seventy per cent of the population live in rural areas, the majority in the great drainage basin regions of the Huang He (Hwang Ho or Yellow River) and the Chang Jiang (Yangtze), where intensive irrigated agriculture produces one-third of the world's rice and also wheat, maize, soya beans and sugar. The country is self-sufficient in cereals, fish and livestock.

Mineral sources are considerable and varied. Oil is extracted from the Yellow Sea oilfields, coal and iron ore in Manchuria and tin, zinc and copper deposits are found in the Yunnan Plateau. China's major exports include petrochemical products, iron and steel, vehicles, cement, fertilizers, textiles and clothing.

In December 1978 Deng Xiaoping launched a second revolution, aimed at transforming and revitalizing the Chinese economy. Dramatic changes have occurred since then. The state has loosened its grip on economic activity to the extent that it now employs only 50 per cent of the working population. Foreign investment, which amounted to $27 billion in 1993, has been encouraged. Land reforms have enabled peasant farmers to lease land and, once they have met their quotas, to sell their produce in local markets. These developments have brought efficiency improvements and a surplus of agricultural labour, which in turn has meant a major migration from the countryside to the booming urban areas.

A major engine for economic progress was the creation in 1979 of Special Economic Zones, where co-operation with foreign investors and independent decision making have been encouraged. The most successful zone, Shenzhen, has increased its population from 320,000 to 2.2 million. Its inhabitants, once among China's poorest, enjoy a per capita income about seven times the Chinese average.

Economic progress has brought with it severe problems of management for the Chinese government. Continuing high levels of economic growth have caused inflation to increase to over 20 per cent. Imports have increased, in particular to support the burgeoning construction industry. Large trade deficits have resulted and tough credit squeezes have been imposed. More recently, the Chinese authorities have admitted that urban unemployment is reaching worrying proportions.

Economic expansion has meant that large sections of the population no longer live in poverty, although the gap in prosperity between coastal and urban areas on the one hand and the interior on the other has widened. One of the keys to future prosperity is population control. Financial incentives to encourage only one child per family have achieved modest success. The government announced that the population had reached 1.2 billion five years earlier than expected and has renewed the one child per family campaign.

ANHUI (ANHWEI)

STATUS	Province
AREA	139,900 sq km (54,000 sq miles)
POPULATION	52,290,000
CAPITAL	Hefei (Hofei)

BEIJING (PEKING)

STATUS	Municipality
AREA	17,800 sq km (6,870 sq miles)
POPULATION	10,870,000
CAPITAL	Beijing

FUJIAN (FUKIEN)

STATUS	Province
AREA	123,100 sq km (47,515 sq miles)
POPULATION	30,610,000
CAPITAL	Fuzhou (Foochow)

GANSU (KANSU)

STATUS	Province
AREA	530,000 sq km (204,580 sq miles)
POPULATION	22,930,000
CAPITAL	Lanzhou (Lanchow)

GUANGDONG (KWANGTUNG)

STATUS	Province
AREA	231,400 sq km (89,320 sq miles)
POPULATION	63,210,000
CAPITAL	Guangzhou (Canton)

GUANGXI-ZHUANG (KWANGSI-CHUANG)

STATUS	Autonomous Region
AREA	220,400 sq km (85,075 sq miles)
POPULATION	42,530,000
CAPITAL	Nanning

GUIZHOU (KWEICHOW)

STATUS . . . . **Province**

AREA . . . . **174,000 sq km**
(67,165 sq miles)

POPULATION **32,730,000**

CAPITAL . . . . **Guiyang (Kweiyang)**

HAINAN

STATUS . . . . **Island Province**

AREA . . . . **34,965 sq km**
(13,500 sq miles)

POPULATION **6,420,000**

CAPITAL . . . . **Haikou**

HEBEI (HOPEI)

STATUS . . . . **Province**

AREA . . . . **202,700 sq km**
(78,240 sq miles)

POPULATION **60,280,000**

CAPITAL . . . . **Shijiazhuang**

HEILONGJIANG (HEILUNGKIANG)

STATUS . . . . **Province**

AREA . . . . **710,000 sq km**
(274,060 sq miles)

POPULATION **34,770,000**

CAPITAL . . . . **Harbin**

HENAN (HONAN)

STATUS . . . . **Province**

AREA . . . . **167,000 sq km**
(64,460 sq miles)

POPULATION **86,140,000**

CAPITAL . . . . . . **Zhengzhou (Chengchow)**

HUBEI (HUPEH)

STATUS . . . . **Province**

AREA . . . . **187,500 sq km**
(72,375 sq miles)

POPULATION **54,760,000**

CAPITAL . . . . **Wuhan**

HUNAN

STATUS . . . . **Province**

AREA . . . . **210,500 sq km**
(81,255 sq miles)

POPULATION **60,600,000**

CAPITAL . . . . **Changsha**

JIANGSU (KIANGSU)

STATUS . . . . **Province**

AREA . . . . **102,200 sq km**
(39,450 sq miles)

POPULATION **68,170,000**

CAPITAL . . . . **Nanjing**

JIANGXI (KIANGSI)

STATUS . . . . **Province**

AREA . . . . **164,800 sq km**
(63,615 sq miles)

POPULATION **38,280,000**

CAPITAL . . . . **Nanchang**

JILIN (KIRIN)

STATUS . . . . **Province**

AREA . . . . **290,000 sq km**
(111,940 sq miles)

POPULATION **25,150,000**

CAPITAL . . . . **Changchun**

LIAONING

STATUS . . . . **Province**

AREA . . . . **230,000 sq km**
(88,780 sq miles)

POPULATION **39,980,000**

CAPITAL . . . . **Shenyang**

NEI MONGOL (INNER MONGOLIA)

STATUS . . . . **Autonomous Region**

AREA . . . . **450,000 sq km**
(173,700 sq miles)

POPULATION **21,110,000**

CAPITAL . . . . **Hohhot**

NINGXIA-HUI (NINGHSIA HUI)

STATUS . . . . **Autonomous Region**

AREA . . . . **170,000 sq km**
(65,620 sq miles)

POPULATION **4,660,000**

CAPITAL . . . . **Yinchuan**

QINGHAI (CHINGHAI)

STATUS . . . . **Province**

AREA . . . . **721,000 sq km**
(278,305 sq miles)

POPULATION **4,430,000**

CAPITAL . . . . **Xining (Yinchuan)**

SHAANXI (SHENSI)

STATUS . . . . **Province**

AREA . . . . **195,800 sq km**
(75,580 sq miles)

POPULATION **32,470,000**

CAPITAL . . . . **Xian (Sian)**

SHANDONG (SHANTUNG)

STATUS . . . . **Province**

AREA . . . . **153,300 sq km**
(59,175 sq miles)

POPULATION **83,430,000**

CAPITAL . . . . **Jinan (Tsinan)**

SHANGHAI

STATUS . . . . **Municipality**

AREA . . . . **5,800 sq km**
(2,240 sq miles)

POPULATION **13,510,000**

CAPITAL . . . . **Shanghai**

SHANXI (SHANSI)

STATUS . . . . **Province**

AREA . . . . **157,100 sq km**
(60,640 sq miles)

POPULATION **28,180,000**

CAPITAL . . . . **Taiyuan**

SICHUAN (SZECHWAN)

STATUS . . . . **Province**

AREA . . . . **569,000 sq km**
(219,635 sq miles)

POPULATION **106,370,000**

CAPITAL . . . . **Chengdu (Chengtu)**

TIANJIN (TIENTSIN)

STATUS . . . . **Municipality**

AREA . . . . **4,000 sq km**
(1,545 sq miles)

POPULATION **8,830,000**

CAPITAL . . . . **Tianjin**

XINJIANG UYGUR (SINKIANG UIGHUR)

STATUS . . . . **Autonomous Region**

AREA . . . . **1,646,800 sq km**
(635,665 sq miles)

POPULATION **15,370,000**

CAPITAL . . . . **Urumqi (Urumchi)**

XIZANG ZIZHIQU (TIBET)

STATUS . . . . **Autonomous Region**

AREA . . . . **1,221,600 sq km**
(471,540 sq miles)

POPULATION **2,220,000**

CAPITAL . . . . **Lhasa**

YUNNAN

STATUS . . . . **Province**

AREA . . . . **436,200 sq km**
(168,375 sq miles)

POPULATION **36,750,000**

CAPITAL . . . . **Kunming**

ZHEJIANG (CHEKIANG)

STATUS . . . . **Province**

AREA . . . . **101,800 sq km**
(39,295 sq miles)

POPULATION **40,840,000**

CAPITAL . . . . **Hangzhou (Hangchow)**

COLOMBIA

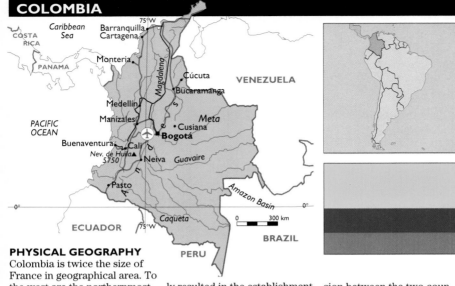

PHYSICAL GEOGRAPHY
Colombia is twice the size of France in geographical area. To the west are the northernmost peaks of the Andes, which are divided into three sub-ranges. Beyond these extend the low plains, with many rivers. Bogotá, the capital, lies on the high plateau east of the Andes. Further east lie the prairies and southwards the jungle of the Amazon.

CLIMATE
Colombia has a tropical climate. Most of the rainfall occurs between March and May and between October and November. Temperatures vary with altitude. Annual rainfall is in the range 799–1,606 mm (32–64 inches).

POLITICS AND RECENT HISTORY
Recent years in Colombia have been dominated by endemic violence, with threats to government control coming from both left wing guerrillas and drug-trafficking cartels.

Colombia, Bolivia and Peru have joined with the USA in an attempt to eliminate the cocaine and heroin trade, but in Colombia the results have been mixed. The highly publicized death of Pablo Escobar, a notorious drug trafficker, mere-ly resulted in the establishment of another cartel based in Cali. Ernesto Samper, a Liberal, who gained the presidency in 1994 with a narrow victory over his Conservative rival, was accused of funding his campaign with money from the drug barons. However, no doubt under strong US pressure, he has recently waged a powerful and moderately successful crackdown on the Cali syndicate.

The guerrilla movement, which is adversely affecting the economy, is still active. The main elements are the Colombian Revolutionary Armed Forces and the National Liberation Army. The latter were responsible for the murder of eight Venezuelan marines in 1995, causing ten-sion between the two countries. President Samper has declared his intention of setting up a peace commission to open dialogue with the rebel factions, in the hope of ending decades of strife.

ECONOMY
Colombia is the world's second largest coffee producer and recent high prices have assisted the nation's economy. Mining is also of great importance and includes coal, gold and emeralds. Colombia's oil wealth has been enhanced by recent discoveries. BP's find at Cusiana is the biggest in the western hemisphere since that at Prudhoe Bay, Alaska, and the government is hoping to earn $3 billion from it by 1997. Such predictions, however, are speculative and may depend on nullifying the threat from guerrilla groups who have in the past sabotaged oil installations.

THE REPUBLIC OF COLOMBIA	
STATUS	Republic
AREA	1,138,915 sq km (439,620 sq miles)
POPULATION	34,520,000
CAPITAL	Bogotá
LANGUAGE	Spanish, Indian languages
RELIGION	95% Roman Catholic, Protestant and Jewish minorities
CURRENCY	Colombian peso (COP)
ORGANIZATIONS	OAS, UN

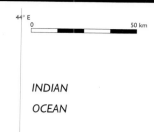

Grande Comore (Njazídja)

Mitsamiouli

Moroni

Mont Kartala 2361

INDIAN OCEAN

44° E

0   50 km

12° S   12° S

Mutsamudu

Sima

Anjouan (Nzwani)

Fomboni

Mohéli (Mwali)

44° E

PHYSICAL GEOGRAPHY
The Comoros consist of three large islands, once forested and of volcanic origin, and some small coral atolls.

CLIMATE
The island group has a climate which is humid all year round, with a moderate rainfall averaging 424 mm (17 inches). It is hot from November to May and drier and slightly cooler for the rest of the year.

POLITICS AND RECENT HISTORY
In 1975, following a referendum the previous year, the three main islands (Grande Comore, Mohéli and Anjouan) unilaterally declared independence from France. The fourth island, Mayotte, where 64 per cent of the population voted against independence, remains under French administration. Over the years, claims to Mayotte by The Comoros have been assessed by various international organizations, including the United Nations. The Comoros republic, too, despite its disagreement over Mayotte, still has close ties with France in the form of a defence pact and aid donations.

However, in 1995 the French government decided to reimpose visa requirements for all Comoros citizens entering Mayotte, fuelling anti-French feelings.

Since independence, the government has been characterized by short-term administrations, coups, shifting allegiances and rivalries between leading families. Political parties have traditionally had a family or sultanate at the core, surrounded by supporters. The current president, Said Mohamed Djohar, was re-elected in 1990 and sought to create a government of national unity. This was ineffective until his party gained an absolute majority in the 1993 legislative elections.

THE FEDERAL ISLAMIC REPUBLIC OF THE COMOROS	
STATUS	Federal Islamic Republic
AREA	1,860 sq km (718 sq miles)
POPULATION	607,000
CAPITAL	Moroni
LANGUAGE	French, Arabic, Comoran
RELIGION	Muslim majority, Christian minority
CURRENCY	Comoro franc (KMF)
ORGANIZATIONS	Arab League, OAU, UN

ECONOMY
The Comoran economy is poor – 80 per cent of the population is rural, mainly engaged in subsistence agriculture. Deforestation has led to soil erosion. Less than half of the land is cultivated, and the country is dependent on imports for food supplies. Despite heavy rainfall, the islands suffer serious water shortages.

The main cash crops are vanilla, where the Comoros are the second largest producer after Madagascar, and cloves, but the value of these commodities is decreasing. The Comoros is also the world's largest producer of ylang ylang, an essence extracted from trees and exported to France for the perfume industry.

There are some prospects for expansion of the fishing industry in Comoran waters and the possibility of developing a tourist industry. However, such endeavours will need great success if they are to bring prosperity to this densely populated republic.

PHYSICAL GEOGRAPHY

The country is for the most part forest or savannah-covered plateaux, drained by the Oubangui and Zaire river systems. Its coast is lined with sand dunes and lagoons.

CLIMATE

Congo has a hot tropical climate with daily temperatures seldom falling below 30°C (86°F). Rainfall is most plentiful in the south of the country, averaging 1,219–1,270 mm (48–50 inches) a year. The dry season runs from June to September.

POLITICS AND RECENT HISTORY

Congo became independent from France in 1960 and was generally known as Congo (Brazzaville) to distinguish it from its neighbour, the former Belgian Congo, which on independence became Congo (Kinshasa) and later Zaire. For most of its history since independence, Congo has been a Marxist state under single-party rule. However, a referendum in March 1992 resulted in an overwhelming majority for multi-party democracy. Later that year, Pascal Lissouba of the Pan African Union of Social Democracy (UPADS) secured the presidency and appointed a coalition government. Although President Lissouba's party has remained dominant, he has failed to gain an adequate working majority. Between 1992 and 1994, there were frequent clashes between the government armed forces and armed supporters of the opposition alliance. In early 1995 a new cabinet took office, but 12 MPs resigned from UPADS to form a new political party.

ECONOMY

The greater proportion of the Congolese population is engaged in agriculture, mainly subsistence farming but also the cultivation of sugar, coffee, cocoa and palm oil for export. The main source of export revenue is, however, oil. Offshore fields produce about eight million tonnes per year, making the country Africa's fifth largest producer. Income is also derived from the timber resources of the forested areas.

Congo is grappling with serious economic problems stemming from a very large foreign debt. Reforms, encouraged by the International Monetary Fund, are necessary. The government has introduced tight fiscal policies to curb public spending and reduce the size of the civil service. It also intends to privatize state assets and hopes to encourage further investment in the oil industry by establishing free zones of oil exploration. Despite this, it has mortgaged further oil revenue to meet short-term debt commitments.

Industrial output has risen, however, and the sugar processing industry is expected to expand. Significant economic development will require investment in the transportation system; the road between Brazzaville and Pointe Noire is currently the nation's only highway.

THE REPUBLIC OF THE CONGO	
STATUS	Republic
AREA	342,000 sq km (132,010 sq miles)
POPULATION	2,443,000
CAPITAL	Brazzaville
LANGUAGE	French, Kongo, Teke, Sanga
RELIGION	50% traditional beliefs, 30% Roman Catholic, Protestant and Muslim minorities
CURRENCY	CFA franc (C Africa) (BEAC)
ORGANIZATIONS	OAU, UN

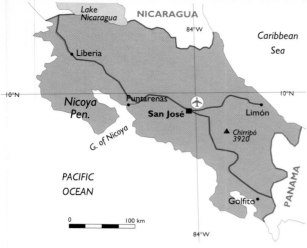

PHYSICAL GEOGRAPHY

Coastal plains, narrow on the Pacific side but broader towards the Caribbean, rise to a central plateau which leads to a range of mountains running the length of the country. Volcanic peaks in this range exceed 3,500 m (11,500 ft).

CLIMATE

Coastal regions experience hot, humid tropical conditions, but the climate on the plateau is more equable. Annual rainfall is in the region 1,793 mm (72 inches). Temperatures range between 19–21°C (66–69°F).

POLITICS AND RECENT HISTORY

An independent nation since 1821, Costa Rica has a history of stable democracy. Indeed, following a minor revolution, the armed forces were disbanded and are constitutionally forbidden. Civil and rural guards maintain security.

The political scene is dominated by two parties, the socialist oriented *Partido de Liberación Nacional* (PLN) and the more conservative *Partido Unidad Social Cristiana* (PUSC). The PLN has generally had the upper hand, but in elections for con-gress and the presidency in February 1990, Rafael Calderón Fournier of the PUSC defeated the PLN candidate. In presidential elections held in February 1994, the PLN candidate, José Figueres, defeated Miguel Rodríguez of the PUSC by only a narrow margin.

Although Costa Rica declared its neutrality in 1983, it has accepted help from the USA to modernize its security forces and has co-operated in the battle against drug trafficking.

ECONOMY

The past few decades have witnessed considerable success for the Costa Rican economy, which has grown consistently. Economic strength is traditionally based on agriculture, but mining and manufacturing (in which food processing dominates) are developing fast and now provide 20 per cent of gross domestic product.

The principal agricultural export and still the biggest earner is bananas, but high-yielding coffee grown on the central plateau is also important, particularly following recent rises in coffee prices. However, government-inspired investment in cash crops has meant that Costa Rica is no longer self-sufficient in traditional foods such as rice, maize and beans, which must now be imported

The tourist trade is developing rapidly, with sharply increasing revenues. Even so, Costa Rica suffers trade balance difficulties and has had to call upon the International Monetary Fund and the World Bank for assistance. This has been provided, subject to requirements that the economy is restructured and privatization programmes introduced. However, the government has expressed its opposition to widespread privatization and has ruled out the handover of some state enterprises. Progress has, in consequence, been limited.

THE REPUBLIC OF COSTA RICA	
STATUS	Republic
AREA	50,900 sq km (19,650 sq miles)
POPULATION	3,005,000
CAPITAL	San José
LANGUAGE	Spanish
RELIGION	95% Roman Catholic
CURRENCY	Costa Rican colon (CRC)
ORGANIZATIONS	CACM, OAS, UN

PHYSICAL GEOGRAPHY
To the south are low undulating plains and rainforest, while the northern areas are savannah plateaux lands.

CLIMATE
Temperatures are warm all year round, between 24°C (75°F) and 27°C (81°F). Annual rainfall, in the range 1,300–2,100 mm (50–84 inches), occurs in two wet seasons near the coast, and a single wet season from June to October in the north.

POLITICS AND RECENT HISTORY
Since independence from France in 1960, Côte d'Ivoire had only one leader, President Félix Houphouët-Boigny. It was feared that his death in December 1993 would bring a constitutional crisis to a nation, hitherto regarded as the most politically stable in Africa. His rule was unopposed and even when multi-party elections were introduced in 1990, he was re-elected with 80 per cent of the vote, conceding only 18 per cent to his rival Laurent Gbagbo.

Under the terms of the constitution, amended in 1990, the parliamentary speaker, Henri Konan Bédié, inherited the presidency until the next elections in 1995. This was challenged by the prime minister, Alassane Ouattara, who resigned rather than serve under Bédié, although they were from the same ruling Democratic Party. Daniel Kablan Duncan was appointed prime minister and Côte d'Ivoire's political stability has been preserved.

ECONOMY
Côte d'Ivoire is the world's leading cocoa producer and Africa's leading producer of coffee. During the 1970s the economy of the country surged, based on the commodity boom. In the late 1980s, however, world prices collapsed and new competition emerged from Malaysian and Indonesian producers, causing some decline in the economy.

Following a necessary and painful 50 per cent devaluation of the CFA franc in early 1994, the economy is doing well. Inflation has been brought under control. Imports have been cut as the country becomes more reliant on its own produce. There remains substantial foreign debt but France has agreed to rescheduling.

The exploitation of offshore reserves of oil and natural gas will prove beneficial to the Ivorian economy, and promises to make the country self-sufficient in these commodities. Further wealth is available from mineral deposits, and Côte d'Ivoire is seeking to encourage foreign investment to assist in their exploitation.

THE REPUBLIC OF CÔTE D'IVOIRE	
STATUS	Republic
AREA	322,465 sq km (124, 470 sq miles)
POPULATION	13,316,000
CAPITAL	Yamoussoukro
LANGUAGE	French, tribal languages
RELIGION	65% traditional beliefs, 23% Muslim, 12% Roman Catholic
CURRENCY	CFA franc (W Africa) (XOF)
ORGANIZATIONS	ECOWAS, OAU, UN

PHYSICAL GEOGRAPHY
In the long Adriatic strip a series of limestone ridges runs parallel to the coast. Central parts around Zagreb are hilly, while in the east Slavonia is a lowland plain.

CLIMATE
There is a contrast in climatic conditions between the Adriatic coast and inland regions. Coastal areas experience Mediterranean conditions and mild winters, with temperatures of between 7–23.5°C (45–73°F) and an average annual rainfall of around 688 mm (27 inches). Inland, rainfall averages 652 mm (26 inches) a year with temperatures of between 0–23.5°C (32–74°F).

POLITICS AND RECENT HISTORY
When Croatia broke away from former Yugoslavia in 1991, the country was plagued by serious conflict between the Serbs and Croats. In January 1992 a United Nations (UN) cease-fire took effect. Croatia gained international recognition and in May 1992 was admitted to the UN. Since then calmer conditions have prevailed, but a continuing threat of instability remains.

Some areas, the largest of which is the territory of Krajina, remained for a long time under ethnic Serb control. The UN Protection Force sought to keep the peace, although the Croatian President wanted their withdrawal. He was persuaded to allow them to stay, albeit in fewer numbers, by a European Union promise of preferential trading links. The futility of the UN position was, however, well demonstrated when Croat forces brushed aside UN forces and drove Serbs from their enclave in western Slavonia. Significantly, President Milosevic in Serbia made no attempt to intervene. Croatia has strengthened its military position against Serb rebels by reaching a formal agreement with both Bosnian Croats and

Muslims, and has largely wrested control from the Serbs within its borders.

In december 1995 Croatia was a signatory to the Dayton peace agreement which guaranteed the independence of Bosnia.

ECONOMY
Croatia has faced severe economic difficulties following independence. It was one of the wealthiest regions of the former Yugoslav republic, with both a thriving agricultural sector and substantial manufacturing industries. However, the loss of markets in Serbia, the effects of war, a reconstruction bill estimated at $20 billion, a budget dominated by defence needs and a severe loss of tourist revenue in Adriatic coastal resorts have left the nation in a difficult position.

Despite these problems, the government intends economic reform. Around 1,000 companies have been privatized, although in reality little cash has been transferred. True economic recovery is dependent upon political stability and an end to armed conflict.

THE REPUBLIC OF CROATIA	
STATUS	Republic
AREA	56,540 sq km (21,825 sq miles)
POPULATION	4,511,000
CAPITAL	Zagreb
LANGUAGE	Serbo-Croat
RELIGION	Roman Catholic majority
CURRENCY	kuna
ORGANIZATIONS	UN

PHYSICAL GEOGRAPHY

Cuba, the largest island in the Caribbean, consists mostly of plains interrupted by three mountain ranges. Forests of pine and mahogany cover 23 per cent of the land.

CLIMATE

The temperature is hot throughout the year, averaging 22–28°C (72–82°F). Rain falls between May and October. Hurricanes sometimes threaten the country in autumn. The average annual rainfall is around 1,194 mm (47 inches).

POLITICS AND RECENT HISTORY

Cuba, one of Communism's remaining strongholds, last held parliamentary elections in February 1993. Fidel Castro retained the presidency; the only way to oppose the re-election of the one-party government was to spoil the ballot paper. The government has, however, announced that in future it will work 'collectively' and will depend less on the authority of a single leader.

ECONOMY

Cuba is facing a crippling economic crisis. The US trade embargo has applied for many years, but until recently the effects of the embargo were compensated by healthy trade with the former USSR. Altogether, 85 per cent of trade was with the former Communist bloc, and consisted of imports of food, machinery and petrol in return for Cuban sugar. The USSR was also generous with aid donations. However, the break-up of the USSR and the collapse of Communism have led to a major disruption of Cuba's trade relations and an end to the aid packages. In 1994 Russia suspended oil shipments because Cuba had not met its quota for the supply of sugar. Cuba has suffered a decline in production, fuel shortages, serious power cuts, and food rationing.

The sugar harvest, which historically accounted for 80 per cent of export earnings, was devastated by storms in 1993, recording a 40 per cent loss on the previous year, and has since declined further.

In the face of economic disaster and in order to control the black market, Cuba legalized the possession and use of hard currency. This is now encouraging dollar remittances from Cubans abroad, and stimulating the trade in the diplomatic dollar shops (where the government raised prices by 50 per cent).

The government has also liberated the economy to the extent of allowing limited private enterprise. This has encouraged some inward investment from the European Union, whose members are inclined to believe that co-operation rather than isolation is the best route to reform. The US government, however, strictly maintains its embargo but has opened direct telephone communication with Havana.

The economic crisis has forced thousands of Cubans to escape by boat to Florida. Faced with a further influx of refugees, the US decreed that potential immigrants picked up at sea would now be returned to Cuba, rather than automatically admitted to the US, as was previously the case.

THE REPUBLIC OF CUBA	
STATUS	Republic
AREA	114,525 sq km (44,205 sq miles)
POPULATION	10,941,000
CAPITAL	Havana (Habana)
LANGUAGE	Spanish
RELIGION	Roman Catholic majority
CURRENCY	Cuban peso (CUP) US dollar (USD)
ORGANIZATIONS	UN

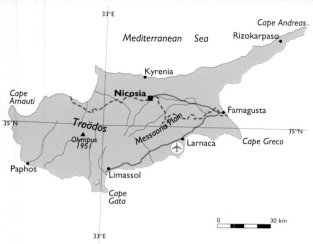

PHYSICAL GEOGRAPHY

The Troödos mountains dominate the centre-west of the island. To the east is the fertile plain of Messaoria, flanked by hills to the northeast.

CLIMATE

The Mediterranean climate gives Cyprus hot summers and mild winters. Rainfall generally occurs in the winter months.

POLITICS AND RECENT HISTORY

Since an invasion by Turkey in 1974, the island of Cyprus has been divided. The self-declared Turkish Republic of Northern Cyprus has been recognized only by Turkey. The United Nations (UN) provide a peace-keeping force to police the 110-mile buffer zone between north and south. Repeated UN sponsored attempts to achieve reunification have foundered, but some co-operative agreements on practical matters give rise to hope that reunification may one day be achieved.

In 1990, Cyprus applied to join the European Union (EU), a move opposed by Rauf Denktash, the Turkish Cypriot leader in the north. However, a timetable towards membership was agreed in early 1995 when Greece lifted its veto on the issue of EU aid to Turkey and the establishment of a Customs Union between the EU and Turkey. The possibility that Cyprus will join the EU before a political settlement for the island has been reached has increased.

ECONOMY

Traditionally, the economy has been based on specialist agricultural products, accounting for 26 per cent of exports, and light manufacturing industries, in particular textiles, clothing, leather and shoes, although manufacturing has suffered some decline.

A growth sector of the economy is the shipping industry, based on favourable tax regimes and Cyprus's strategic position between Asia and Europe. A total of 1,700 ocean-going vessels now fly the Cyprus flag, and over 100 companies are in shipping services. Recently, Cyprus has also increased in popularity as an international business centre and a home for offshore business registrations. However, the vital tourist industry, which welcomed two million arrivals in 1992, has contracted because of recession.

In northern Cyprus much of the region's main export, citrus fruits, are produced by concerns originally owned by Asil Nadir (who was charged by the UK authorities with false accounting and absconded to northern Cyprus). Although Mr Nadir at first enjoyed the support of the northern Cyprus government, he has since incurred their displeasure for failure to pay taxes.

THE REPUBLIC OF CYPRUS	
STATUS	Republic
AREA	9,250 sq km (3,570 sq miles)
POPULATION	726,000
CAPITAL	Nicosia
LANGUAGE	Greek, Turkish, English
RELIGION	Greek Orthodox majority, Muslim minority
CURRENCY	Cyprus pound (CYP)
ORGANIZATIONS	Comm., Council of Europe, UN

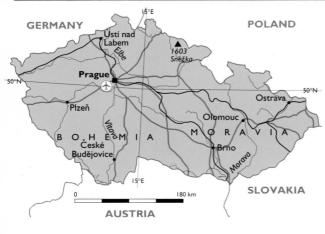

PHYSICAL GEOGRAPHY

The Czech Republic, in the heart of Europe, is a land of rolling countryside, wooded hills and fertile valleys. In Bohemia, to the west, the upper basin of the River Elbe system, in the centre of which lies Prague, is surrounded by mountains. Moravia, separated from Bohemia by the Moravian Heights, is a lowland area centred on the town of Brno.

CLIMATE

The climate is temperate but with continental characteristics, featuring warm summers and relatively cold winters. Annual rainfall ranges between 483–525 mm (19–21 inches).

POLITICS AND RECENT HISTORY

The former republic of Czechoslovakia freed itself from Soviet satellite status in 1989 and in 1990 the popular liberal playwright, Vaclav Havel, was elected president. Although that transition was managed without undue turbulence, stresses developed between the majority Czech and the minority Slovak peoples. Nationalistic forces on the Slovak side forced a separation into the Czech and Slovak

Republics, which came into being on 1 January 1993.

The Czech prime minister, Vaclav Klaus of the Civic Democratic Party, leads a four-party coalition. The government has pursued a policy of strengthening relations with western Europe and inherited Associate status with the European Union (EU). Full membership of the EU is predicted for the turn of the century. Amid political change, President Havel is now a figurehead, although he still commands widespread popularity.

ECONOMY

The Czech Republic was an industrial powerhouse of the Soviet empire and has been able to manage its economic independence with more suc-

THE CZECH REPUBLIC	
STATUS	Federal Republic
AREA	78,864 sq km (30,449 sq miles)
POPULATION	10,333,000
CAPITAL	Prague (Praha)
LANGUAGE	Czech
RELIGION	40% Roman Catholic, 55% no stated religion
CURRENCY	Czech crown or koruna (CEK)
ORGANIZATIONS	Council of Europe, OECD, UN

cess than its neighbours. It possesses valuable raw materials (coal, minerals and timber) and has a steel industry using cheap iron ore from the Ukraine. Although known for its production of cars, aircraft, tramways and locomotive diesel engines, traditionally the region has specialized in arms manufacture. However, demand for arms collapsed with the end of the West–East arms race, and the industry is now diversifying to suit new customers.

The most dramatic change has been the transformation from state ownership to privatization by means of a voucher system. Almost 80 per cent of industry and commerce is in private hands, making the country the world's largest share-owning democracy.

The economy is showing strength as it emerges from recession. The currency is stable and reaching convertibility. Inflation is falling towards 3 per cent and unemployment stands at a low 3 per cent. It is not surprising that foreign investors, particularly from Germany, are attracted.

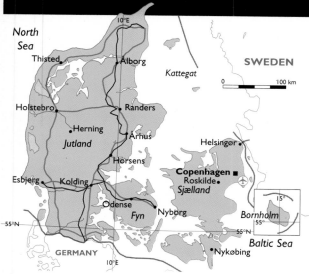

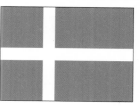

PHYSICAL GEOGRAPHY

Denmark consists of the Jutland peninsula and numerous small islands. The country is low-lying, with a mixture of fertile and sandy soils, generally of glacial origin.

CLIMATE

The climate is cool and temperate with rainfall spread fairly evenly throughout the year. Copenhagen has an average summer temperature of 20°C (68°F), with 0°C (32°F) in winter.

GOVERNMENT AND RECENT HISTORY

Elections to the Danish parliament, the *Folketing*, are by proportional representation which has generally resulted in coalition governments. In recent years the Conservatives have shared government with three minor parties. However, in January 1993 one of these parties, the Radical Liberals, split with the Conservatives and switched allegiance to the Social Democratic Party. A new coalition was formed, led by Poul Nyrup Rasmussen of the Social Democratic Party. In the 1994 general election, Rasmussen retained power in a coalition between his own party, the Centre Democrats, and the Radical Liberals.

Danish politics have been dominated in recent times by its membership of the European Union and in particular the provisions of the Maastricht Treaty over which the nation remains divided.

ECONOMY

The traditional basis of the Danish economy is agriculture, in particular dairy products and bacon, much of which is exported to the UK and Germany. Oil and gas, beer, sugar and fishing also contribute to the economy and further recent oil discoveries in the Danish sector of the North Sea assure self-sufficiency well into the next century.

Manufacturing, developed in association with agriculture, accounts for 75 per cent of exports and includes machinery needed for the dairy industry and pharmaceutical products.

The effects of recession have caused some damage to the Danish economy. Gross domestic product has fallen and unemployment has risen sharply to 12 per cent, although inflation remains low. This is in part due to the high interest rates which have allowed the Danish krone to remain relatively stable within the European exchange rate mechanism, to which Denmark is committed.

In September 1993 the government introduced reflationary budget proposals, involving a 7 per cent increase in spending, with the object of reducing unemployment, the signs of post-recession recovery are now evident.

One important development is the planned bridge and tunnel linking Denmark and Sweden, which has been approved by both governments.

THE KINGDOM OF DENMARK	
STATUS	Kingdom
AREA	43,075 sq km (16,625 sq miles)
POPULATION	5,212,000
CAPITAL	Copenhagen (København)
LANGUAGE	Danish
RELIGION	94% Lutheran, Protestant and Roman Catholic minority
CURRENCY	Danish krone (DKK)
ORGANIZATIONS	Council of Europe, EEA, EU, NATO, OECD, UN

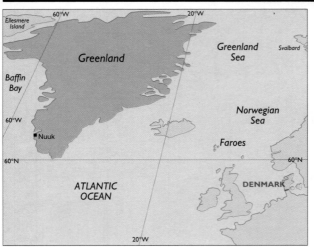

GREENLAND	
STATUS	Self-governing Island Region of Denmark
AREA	2,175,600 sq km (839,780 sq miles)
POPULATION	57,000
CAPITAL	Nuuk (Godthåb)

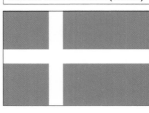

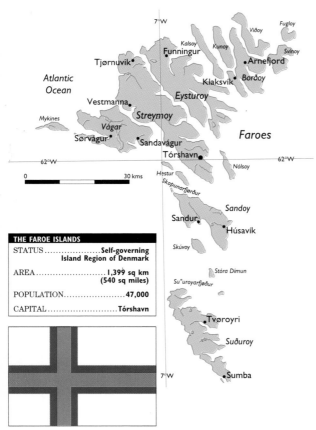

THE FAROE ISLANDS	
STATUS	Self-governing Island Region of Denmark
AREA	1,399 sq km (540 sq miles)
POPULATION	47,000
CAPITAL	Tórshavn

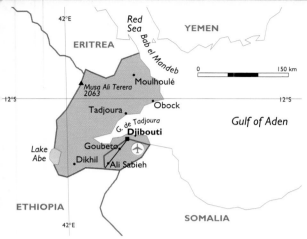

PHYSICAL GEOGRAPHY

Djibouti consists almost entirely of low-lying desert, with
some areas below sea-level. To
the north lies a mountainous
area where the highest point,
Musa ali Terara, reaches 2,062
m (6,768 ft).

CLIMATE

Temperatures are hot all year,
between 25°–35°C (78°F–96°F),
although it is cooler between
October and April. The annual
rainfall is less than 130 mm (5
inches).

POLITICS AND RECENT HISTORY

Djibouti became independent
from its former status as the
French Territory of the Afars
and Issas in 1977. Before 1967
it was known as French
Somaliland.

The country was officially a
one-party state from 1981 until
1992, when a new constitution
permitted the formation of no
more than four political par-
ies. In the 18 years since independence, Hassan Gouled
Aptidon, believed to be in his
late eighties, has been president. His dictatorship has been
unopposed, although in the last
elections held in May 1993, he
and the People's Progress

Assembly claimed a clear
majority despite less than half
the electorate having voted.

Djibouti has a history of
clashes between the two main
ethnic groups, the Afars (or
Danakil) and the Somali Issas.
For three years, up to the end
of 1994, an armed Afar opposition group, the Front for the
Restoration of Unity and
Democracy (FRUD), had been
fighting the government in
protest at political domination
by the Issas. On 26 December
1994 a peace agreement was
signed between the government and FRUD, involving revision of the constitutional and
electoral lists.

France, the former colonial
power, exerts a major influence
in the country, supporting it

financially and maintaining a
naval base and a garrison of
4,000 troops. It, however,
refrained from intervening in
the civil war.

ECONOMY

The land of Djibouti is barren,
and the economy relies on the
country's major asset – its deep
natural port and its strategic
position along a shipping route.
During the Gulf crisis in
1990–1, Djibouti was a staging
post for the French contingent
en route to join the US-led
forces in Saudi Arabia, and it
served as a base for United
Nations operations in Ethiopia
and Somalia. The USA has provided aid in return for the use
of Djibouti's port and airfield
facilities. There is a rail link to
Addis Ababa, and Djibouti's
port was vital to Ethiopia when
that country was cut off from
the ports in Eritrea.

Banking and transport services are economically important, and this is the sector
believed to have greatest
potential for expansion.

THE REPUBLIC OF DJIBOUTI	
STATUS	Republic
AREA	23,000 sq km (8,800 sq miles)
POPULATION	557,000
CAPITAL	Djibouti
LANGUAGE	French, Somali, Dankali, Arabic
RELIGION	Muslim majority, Roman Catholic minority
CURRENCY	Djibouti franc (DJF)
ORGANIZATIONS	Arab League, OAU, UN

Dominica Passage
61°15'W
Vieille Case
Portsmouth
Wesley
Marigot
Morne Diablotins 1432
Morne Raquette
St. Joseph
Rosalie
Morne Trois Pitons 1387
Caribbean Sea
Roseau
15°15'N
15°15'N
Soufrière
Berekua
Martinique Passage
61°15'W
ATLANTIC OCEAN
0   20 km

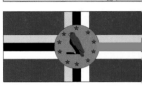

PHYSICAL GEOGRAPHY
Dominica, the largest of the Windward Islands, is mountainous and forested with a coastline of steep cliffs. Evidence of its volcanic origin is provided by the presence of geysers and hot springs.

CLIMATE
The climate is tropical, with average temperatures exceeding 25°C (77°F) and abundant rainfall, averaging 1,750 mm (68 inches) a year in the coastal regions, and up to 6,250 mm (246 inches) in the mountains, especially in the summer months.

POLITICS AND RECENT HISTORY
Dominica, a former British colony, became a fully independent member of the Commonwealth in 1978, after 11 years of self-government. The President, Crispin Sorhaindo, is head of state and appoints the prime minister and cabinet. At the first elections following independence, held in 1980, Dame Mary Eugenia Charles of the centre-right Dominica Freedom Party defeated her Labour Party opponent, Patrick John. She maintained her grip on power in two subsequent elections,

despite attempted coups. In 1983, as head of the Organization of Eastern Caribbean States, she was instrumental in the involvement of the USA in leading the invasion of Grenada. However, in the June 1995 elections, the Labour Party and the Freedom Party, now led by Brian Alleyne, were defeated by Edison James' United Workers Party and he was sworn in as prime minister.

In recent years Dominica's government has been among the most active proponents of economic and political union of the Windward Islands, but has had to contend with traditional inter-island political rivalries.

ECONOMY
In common with its Caribbean

neighbours, the Dominican economy is highly dependent upon its agricultural output. The region is vulnerable to adverse weather conditions and there is always the danger of hurricanes, which have hindered production in the past. Bananas, the most important crop, occupy about a quarter of cultivated land and account for half of the island's exports. About 40 per cent of the total land is forested, and there is a controlled timber industry. There are small but developing coffee and cocoa production sectors. Citrus fruits, especially oranges and grapefruits, are exported to the UK and USA. Coconuts are grown, and coconut oil is used in a domestic industry which has shown increased profitability.

On account of the black volcanic sandy beaches, tourism is undeveloped compared with other Caribbean islands. Dominica's chief attractions are its scenery and natural history. However, custom is being stimulated by a bay-front project, opened in August 1993 and financed by the British government, which has announced a further £5 million for additional development

THE COMMONWEALTH OF DOMINICA	
STATUS	Commonwealth State
AREA	751 sq km (290 sq miles)
POPULATION	72,000
CAPITAL	Roseau
LANGUAGE	English, French patois
RELIGION	80% Roman Catholic
CURRENCY	E Caribbean dollar (XCD) (also French franc and £ sterling)
ORGANIZATIONS	Comm., OAS, UN

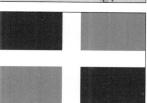

PHYSICAL GEOGRAPHY
The country is one of relatively high mountains, fertile valleys and an extensive coastal plain in the east. The highest peaks in the Cordillera Central exceed 3,000 m (9,800 ft).

CLIMATE
The Dominican Republic has a hot tropical climate with average temperatures of around 25°C (77°F). Rainfall is heavy, especially in the summer months.

POLITICS AND RECENT HISTORY
The Dominican Republic has had civilian rule since the end of military dictatorship in 1966. For much of this time the presidency has been held by Joaquín Balaguer and his right wing party the *Partido Reformista Social Cristiano*. In 1994, President Balaguer gained a seventh term of office in a very narrow victory over Peña Gómez of the left wing *Partido Revolucionario Dominicano*. The election was widely regarded as corrupt and Balaguer, blind and crippled, agreed with his opponent to a further election in late 1995. The National Assembly has since postponed this until 1996 and also legislated against consecutive terms of office.

ECONOMY
The economic wealth of the country is based on primary production in agriculture and mining. However, the traditional dependence on sugar has diminished with the progressive decline in world sugar prices. Other established crops such as coffee, rice and tobacco, have also suffered, but newer products, including cocoa, fruit and vegetables, have shown improvement. Gold, silver, nickel, bauxite and other minerals account for over 40 per cent of the republic's merchandise exports, although returns from nickel and bauxite have fallen with a weakening in world prices.

Some of the shortfall has been offset by a burgeoning tourist trade. Tourism is centred on the north coast Puerto Plata resorts, and most visitors are from the USA and Canada. In recent years the area has also become a popular destination for package holidays from the UK, and between 1987 and 1990 tourist figures soared from 5,000 to 65,000 a year.

Despite the improved income from tourism, the balance of trade has worsened significantly in recent years; exports have dropped while imports have risen to record levels. The International Monetary Fund has provided support in the form of stand-by agreements associated with economic reform. In 1995 the government ended subsidies to over 30 state-owned companies.

President Balaguer incurred some US displeasure over his opposition to the reinstatement of President Aristide in neighbouring Haiti and his lukewarm attitude towards the UN sanctions imposed on that country.

THE DOMINICAN REPUBLIC	
STATUS	Republic
AREA	48,440 sq km (18,700 sq miles)
POPULATION	7,608,000
CAPITAL	Santo Domingo
LANGUAGE	Spanish
RELIGION	90% Roman Catholic, Protestant and Jewish minorities
CURRENCY	Dominican peso (DOP)
ORGANIZATIONS	OAS, UN

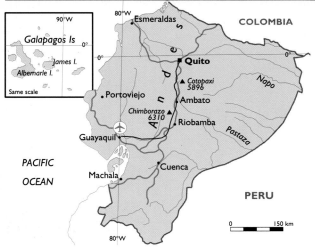

PHYSICAL GEOGRAPHY

Ecuador comprises three distinct physical zones – a broad coastal plain, the high ranges of the Andes where Mt. Chimborazo climbs to well over 6,000 m (20,000 ft) and the forested upper Amazon basin to the east. The territory also includes the Galapagos islands, nearly 1,000 km (620 miles) to the west.

CLIMATE

The climate is tropical, with hot moist conditions along the coast and east of the Andes, and temperate mountainous regions.

POLITICS AND RECENT HISTORY

Ecuador emerged from military dictatorship in 1979. The military administrations were never as severe or repressive as in other parts of Latin America and the armed forces are well regarded by the population. It is likely that a military candidate, General José Gallardo, will stand in the next presidential elections. In 1992, President Durán Ballén of the right wing Republican Unity Party was elected, replacing the left wing administration of Rodrigo Borja, whose austerity

measures had not found favour with the population.

Ecuador has a long-running boundary dispute with Peru, which led to an outbreak of fighting in 1995. A cease-fire was agreed, but the issue remains unresolved.

ECONOMY

Ecuador's economic performance is closely linked to oil. Before exploitation of the commodity in the 1970s, cocoa and coffee provided 83 per cent of export revenue. By 1992 these crops were worth 25 per cent of revenue and oil 44 per cent. During the 1980s oil production was affected by the collapse of world prices and, in 1987, by a severe earthquake, but by the 1990s was recovering. President Ballen left the Organization of Petroleum Exporting Countries

THE REPUBLIC OF ECUADOR	
STATUS	Republic
AREA	461,475 sq km (178,130 sq miles)
POPULATION	11,221,000
CAPITAL	Quito
LANGUAGE	Spanish, Quechua, other Indian languages
RELIGION	90% Roman Catholic
CURRENCY	sucre (ECS)
ORGANIZATIONS	OAS, UN

in November 1992, allowing the country to produce as much oil as it wishes without external constraint.

The government has encountered opposition from the indigenous Indian population, who have complained of water pollution caused by oil exploitation and, more recently, over land redistribution programmes.

Agriculture is still very important to the economy, and is the largest employer. Ecuador is the world's leading exporter of bananas, and these and shrimps are now ahead of coffee and cocoa in importance. However, the shrimp industry has faced continuing problems due to water pollution, caused, ironically, by banana fungicides.

The government is committed to a free market economy and has embarked on a major privatization programme. There has, however, been little progress because Congress is controlled by the opposition. The government has also taken the country into the Andean Free Trade Area.

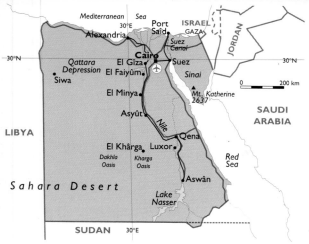

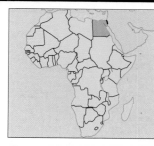

PHYSICAL GEOGRAPHY
Except for the Nile valley, Egypt is desert and semi-desert. The most rugged parts are in the Sinai peninsula, where Mt. Katherine reaches 2,637 m (8,650 ft). The western desert, mostly plateau, falls to altitudes below sea-level.

CLIMATE
The climate is hot in summer and mild in winter. Rainfall is negligible everywhere, although a little occurs in winter along the Mediterranean coast.

POLITICS AND RECENT HISTORY
Egypt is essentially a one-party state which invests great power in the presidency. Hosni Mubarak, who came to power after the assassination of Anwar Sadat in 1981, was re-elected, unopposed, for a third six-year term in October 1993. His supporters, the National Democratic Party, retained control of parliament in the last general election of 1990, although these were boycotted by most opposition parties.

The government is plagued by a recent upsurge of Islamic fundamentalism. The Moslem Brotherhood, an international organization banned but gener-ally tolerated in Egypt, espouses the democratic process to advance its cause. The government has clamped down hard on more radical groups, prompting accusations of human rights abuses and increasing the government's traditional unpopularity. Internationally, President Mubarak has fared much better and Egypt retains its position as the leading force in the Arab world.

ECONOMY
The Egyptian economy is much affected by demographic factors. Population growth in the already over-populated country is 27 per 1,000 per annum. This has led to a migration from crowded rural areas into even more crowded towns. As much as 20 per cent of the population is unemployed. Economic reform through privatization and stimulation of the private sector is slow, hampered by excessive bureaucracy, although some progress has been made. A privately owned oil refinery has been opened and state-owned textile enterprises are to be privatized through public subscription.

The Suez Canal is always dominant in Egypt's affairs. A recent decline in the numbers of ships passing through the canal has been compensated for by raising tolls. Widening and deepening of the canal is being undertaken to counteract the expansion of the rival Sumed oil pipeline, which will take oil between the Red Sea and the Mediterranean.

Tourism was seriously damaged in 1993, in the wake of violence by radical fundamentalist groups. More recently, industry has shown signs of recovery. Exploitation of natural gas is expected to provide a boost to the economy and it is likely that some will be exported via a pipeline to Israel. Despite some positive aspects, the economy is somewhat precarious and continues to be dependent on US aid, which now stands at $4 billion a year.

THE ARAB REPUBLIC OF EGYPT	
STATUS	Republic
AREA	1,000,250 sq km (386,095 sq miles)
POPULATION	57,673,000
CAPITAL	Cairo (El Qâhira)
LANGUAGE	Arabic, Berber, Nubian, English, French
RELIGION	80% Muslim (mostly Sunni), Coptic Christian minority
CURRENCY	Egyptian pound (EGP)
ORGANIZATIONS	Arab League, OAU, UN

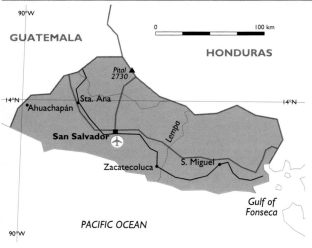

PHYSICAL GEOGRAPHY

El Salvador features a Pacific coastal plain and inland mountain ranges of volcanic origin, the highest peaks of which exceed 2,000 m (6,500 ft).

CLIMATE

The coastal plain is hot, with heavy summer rainfall, whereas the highlands enjoy a cooler temperate climate. Temperatures range between 24–26°C (75–79°F), with an average annual rainfall of 1,780 mm (70 inches).

POLITICS AND RECENT HISTORY

For 12 years until 1992, El Salvador was torn by civil war. The left wing Farabundo Martí National Liberation Front (FMLN) waged guerrilla warfare against the government, which received substantial support from the USA. In the 1988 presidential elections, Alfredo Cristiani of the Republican National Alliance Party (ARENA) defeated the more moderate Christian Democrat incumbent, José Duarte. Although President Cristiani had been seen as the mouthpiece of ARENA's late leader, the far-right Major Roberto D'Aubuisson, he in fact set out to achieve peace. Through the mediating efforts of the United Nations (UN) a peace agreement and cease-fire was agreed with the FMLN in 1992. By and large the fragile peace held and democratic elections to the parliament secured victory for ARENA and strengthened its hold on power.

UN investigations into the activities of the government during the civil war have revealed complicity by the army in atrocities. Under pressure from the UN and the US government, President Cristiani was forced to dismiss a number of senior officers, including the defence minister. Meanwhile, the left wing movement has given up its arms caches and become a legal political party. Nevertheless, ARENA was able to consolidate its hold on power when, in the 1994 presidential elections, its candidate Armando Calderón Sol won by a substantial margin over his centre-left opponent Ruben Zamora.

ECONOMY

Inevitably, El Salvador's economy suffered from the ravages of civil war. However, there are now signs of recovery in the important agricultural sector, where coffee is the most valuable crop, providing about 50 per cent of the country's export revenue.

Manufacturing, which includes footwear, textiles and pharmaceuticals, is well developed, but even so foreign aid is necessary to assist recovery. The International Monetary Fund has, for example, approved a stand-by loan of $4.9 million but, as a condition, requires economic reform designed to enhance joint commercial and investment activities.

THE REPUBLIC OF EL SALVADOR	
STATUS	Republic
AREA	21,395 sq km (8,260 sq miles)
POPULATION	5,517,000
CAPITAL	San Salvador
LANGUAGE	Spanish
RELIGION	80% Roman Catholic
CURRENCY	El Salvadorean colon (SVC)
ORGANIZATIONS	CACM, OAS, UN

PHYSICAL GEOGRAPHY
Equatorial Guinea consists of the island of Bioko and the mainland region of Rio Muni.

CLIMATE
The hot and humid climate has heavy rainfall all year round.

POLITICS AND RECENT HISTORY
Equatorial Guinea gained independence from Spain in 1968. In 1982 the country's constitution was revised, after United

THE REPUBLIC OF EQUATORIAL GUINEA	
STATUS	Republic
AREA	28,050 sq km (10,825 sq miles)
POPULATION	379,000
CAPITAL	Malabo
LANGUAGE	85% Fang, Spanish, Bubi, other tribal languages
RELIGION	96% Roman Catholic, 4% animist
CURRENCY	CFA franc (C Africa) (XAF)
ORGANIZATIONS	OAU, UN

Nations pressure, but accusations of human rights abuses continued. A further constitution was drafted in 1991, incorporating principles of multiparty democracy. However, reports indicated that security forces were torturing and killing suspected dissidents.

The first elections, held in November 1993, were won by President Obiang's ruling party. However, there was no electoral roll, leaders of some opposition parties were prohibited from campaigning and all media reporting was quashed. Seven of the 14 parties united, campaigning for an election boycott, and about a third of the electorate voted.

ECONOMY
The island of Bioko, despite its smaller size, generates most of the country's income. Primary exports are timber and cocoa, since coffee production has

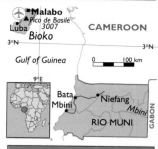

substantially declined. An oil production and exploration industry, begun in 1992-3, has the potential to transform the economy, but is more likely to reinforce corrupt government.

PHYSICAL GEOGRAPHY
Northern regions are an extension of the Ethiopian high plateau. Southwards, the coastal plain forms part of the Great East African Rift.

CLIMATE
The coast is hot but inland temperatures are much lower. Rainfall is unreliable.

POLITICS AND RECENT HISTORY
Following Italian and British colonial rule, Eritrea was incorporated into Ethiopia in 1952.

THE STATE OF ERITREA	
STATUS	Republic
AREA	91,600 sq km (35,370 sq miles)
POPULATION	3,345,000
CAPITAL	Asmara
LANGUAGE	Arabic, native languages, English
RELIGION	50% Christian, 50% Muslim
CURRENCY	Ethiopian birr
ORGANIZATIONS	OAU, UN

On the fall of the Mengistu regime in Ethiopia, the Eritrean People's Liberation Front (renamed the People's Front for Democracy and Justice) declared *de facto* secession. Independence, approved by referendum, was gained in1993. A transitional government has set up the National Eritrean Council, chaired by the president Isaias Afwerki. Although Eritrea is still a one-party state, a constitution is being drafted for elections in 1997, and will allow other political parties to compete.

ECONOMY
The main objective has been to improve agriculture and reduce food aid. Terraces and roads have been rebuilt, and saplings have been planted to combat erosion, bringing an extra 25 per cent of land into cultivation. Poor harvests have hindered progress and food aid has been required. Eritrea joined the World Bank and International

Monetary Fund in 1994, from whom assistance will be forthcoming. Even so, the country has shown a self-reliant attitude and is, for example, reconstructing the Massawa–Asmara railway from its own resources.

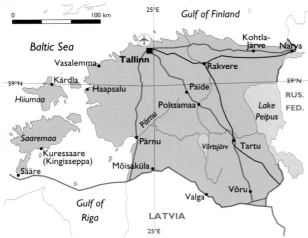

PHYSICAL GEOGRAPHY
Estonia's flat or undulating landscape, one-third of which is forested, is generally low-lying with numerous lakes. There are about 800 islands.

CLIMATE
The climate is temperate with warm summers, cold winters and a heavy rainfall of 500–700 mm (20–28 inches) evenly distributed throughout the year.

POLITICS AND RECENT HISTORY
In August 1991, while the abortive coup was taking place in Moscow, the Estonian parliament declared independence. This was recognized by the former USSR two weeks later.

Estonia led the way in the USSR towards separation. To this end, nationalist groups collaborated with the Communist Party and local laws progressively took precedence over Soviet legislation. Shortly after independence, Estonian was restored as the country's official language and a law was introduced whereby competence in the language is a prerequisite to obtaining nationality.

In July 1992 Estonia introduced its own currency, the kroon, together with a new constitution which included provision for a national assembly to elect a president. In October of that year, Lennart Meri was sworn in, replacing Arnold Rüütel, formerly Chairman of the Supreme Council. Since independence, government has for most of the time been in the hands of the Fatherland Party, under its dynamic leader Maat Laar, who is widely regarded as being responsible for Estonia's success. However, he was defeated in a vote of confidence, for engaging in too much secrecy. Subsequently, in early 1995, the Fatherland Party was decisively defeated in elections. The Coalition Party emerged as the largest single party.

THE REPUBLIC OF ESTONIA	
STATUS	Republic
AREA	45,100 sq km (17,413 sq miles)
POPULATION	1,507,000
CAPITAL	Tallinn
LANGUAGE	Estonian, Russian
RELIGION	Lutheran, Roman Catholic
CURRENCY	kroon (EKR)
ORGANIZATIONS	Council of Europe, UN

ECONOMY
Apart from its forests and rich oil-shale deposits, there are few natural resources in Estonia. Industries include timber, furniture production, shipbuilding, leather, fur and food processing. Livestock and dairying are important.

In October 1991 a land reform law permitted the return of land to former owners, and 4,000 farms have reverted to private ownership.

Estonia's transition to a market economy has met with astounding success. The policy of pegging the kroon to the Deutschmark has paid off. Inflation has been rapidly brought under control and the economy is unquestionably the most stable among the former Soviet Republics. Estonia now has Associate status with the European Union.

Despite the change of government, the reforming and successful financial policies are likely to be maintained.

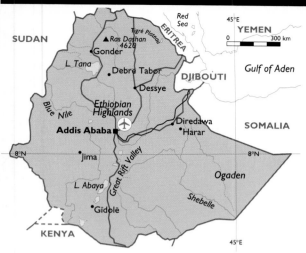

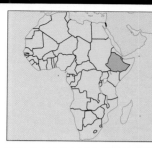

PHYSICAL GEOGRAPHY
Western Ethiopia, including the Tigré Plateau and the Semien Mountains, is a mountainous region of mainly volcanic origin traversed from northeast to southwest by the Great Rift valley. Eastern Ethiopia is mostly arid plateaux country.

CLIMATE
Highland regions have pleasant warm climates and, although droughts occur, in normal years rain falls during the summer months in appreciable quantities. Eastern Ethiopia is hot and generally dry.

POLITICS AND RECENT HISTORY
In 1974 Emperor Haile Selassie was deposed by a military coup. Military rule continued for a further 13 years until 1987, when a new constitution provided for civilian rule. However, the previous military leader, Mengistu Haile Mariam, became president. His Marxist regime was overthrown in May 1991 and the Ethiopian People's Revolutionary Democratic Front (EPRDF) formed a transitional government. The EPRDF is an association of various opposition groups in which the Tigrean

people are dominant. Ethiopia has developed a new constitution on federal lines, designed to give greater power to regions, defined according to their racial composition. In 1995 elections were held on the basis of this constitution and resulted in a resounding victory for the EPRDF, although some opposition parties boycotted the election.

ECONOMY
Ten years ago Ethiopia suffered from continuing civil war and famine in which thousands died. Economic progress since then has been such that the threat of starvation on the same scale has receded. Agriculture has improved dramatically and a record harvest of grains and pulses has been achieved. There are still areas

of malnutrition and considerable food aid, measured at 600,000 tonnes, is necessary.

The Ethiopian government is moving towards its declared aim of self-sufficiency, partly through steps towards a market economy and a focus on agriculture. The collective system has been dismantled and land returned to peasant farmers, who are showing greater interest in conservation and good farming methods.

With food prices stable, the government has been able to exert control over inflation. It has reduced military spending and encouraged some foreign investment. The economy has benefited from increases in the world price of coffee, which accounts for more than half of Ethiopia's export revenue.

THE REPUBLIC OF ETHIOPIA	
STATUS	Republic
AREA	1,104,300 sq km (426,370 sq miles)
POPULATION	51,859,000
CAPITAL	Addis Ababa (Adis Abeba)
LANGUAGE	Amharic, English, Arabic
RELIGION	Ethiopian Orthodox, Muslim, animist
CURRENCY	birr (ETB)
ORGANIZATIONS	OAU, UN

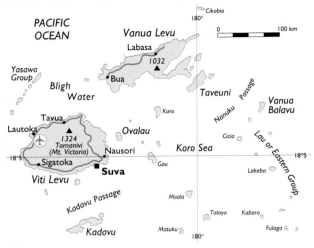

PHYSICAL GEOGRAPHY

Fiji comprises over 300 islands, ranging from tiny atolls to the two largest, Vanua Levu and Viti Levu, which are of volcanic origin and mountainous. The smaller islands mainly comprise coral reefs. The highest point, Tomanivi, on Viti Levu, reaches 1,324 m (4,344 ft).

CLIMATE

The climate is tropical, with a rainy season from December to April.

POLITICS AND RECENT HISTORY

In 1970 Fiji became independent from the UK and joined the Commonwealth. Since then racial tension and rivalry have increased between the native Fijian population and the Indian population, the latter descendants of labour brought in by the British in the late 19th century to work in the sugar plantations. During 1987 the first ever Indian-majority government was elected, but it was overthrown by the military, led by Colonel Sitiveni Rabuka. Fiji was declared a republic and its Commonwealth membership lapsed. Civilian rule was reinstated

in 1990 with a new constitution which guaranteed a parliamentary majority for ethnic Fijians, as well as providing conditions where Fijians would receive preferential access to jobs in the public sector. In May 1992 elections, Rabuka, now a civilian, became prime minister and in December he proposed a government of national unity, a coalition with the Indian-dominated opposition. This was greeted by much debate and, after disagreement over other government issues, the Fiji Labour Party withdrew from parliament in 1993, making the prospect of political unity unlikely. At further elections in 1994, Rabuka retained power, but it was subsequently announced that the constitution will be reviewed.

There is speculation that

THE REPUBLIC OF FIJI	
STATUS	Republic
AREA	18,330 sq km (7,075 sq miles)
POPULATION	771,000
CAPITAL	Suva
LANGUAGE	Fijian, Hindi, English
RELIGION	51% Methodist Christian, 40% Hindu, 8% Muslim
CURRENCY	Fiji dollar (FJD)
ORGANIZATIONS	Col. Plan, UN

Fiji may seek to rejoin the Commonwealth, a move that India would be certain to veto, unless a constitution fairer to the Indian population is established.

ECONOMY

The main product is sugar, providing 45 per cent of export revenue. Fiji presently has quota arrangements with the European Union (which takes about 35 per cent of its sugar) and the USA, allowing it to sell sugar at twice the world price. For further protection, Fiji announced in 1993 that it was applying for membership of the General Agreement on Tariffs and Trade.

Gold mining and tourism are the other key foreign exchange earners, while timber and fish, especially tuna, are increasing in importance.

A major issue in the future of Fiji's economy is the stability of the populace. The Fijians generally own the land, but the Indian population dominates the professions and commerce and contributes 80 per cent of the taxes. An increasing emigration of Indians in recent years must take its toll.

PHYSICAL GEOGRAPHY
In Finland forests cover 70 per cent of the land area and water another 10 per cent. The Saimaa Lake area is Europe's largest inland water system.

CLIMATE
Summers are short but quite warm. Winters are long and severe and in the north the days are sunless. The sea freezes for several miles out from the coast during the severe winters.

POLITICS AND RECENT HISTORY
Finland gained independence from Russia in the 1917 revolution, but has for most of its history since then, occupied a political position of neutrality between the Soviet bloc and Western Europe. This situation changed at the beginning of 1995, when Finland joined the European Union (EU) after 57 per cent of the electorate had voted in favour of membership. At general elections held within weeks of EU membership, the Social Democratic Party, under its leader Paavo Lipponen, won the largest number of seats. It formed a coalition government with four other parties, but excluded the Centre Party, which had led the previous administration.

ECONOMY
Forestry products (timber, wood pulp and paper), which in 1980 comprised 80 per cent of Finnish exports, now account for 40 per cent of the export total, and engineering, in particular shipbuilding and forest machinery, make up another 35 per cent. Finland is

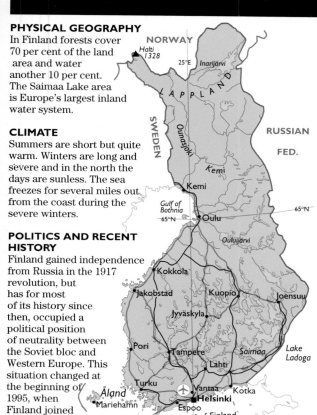

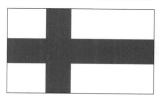

virtually self-sufficient in basic foodstuffs such as dairy products, grains and root crops. The country depends heavily on energy imports, producing only 30 per cent of its total consumption.

Finland's recent economic history has been turbulent. It has run high budget deficits and its economy suffered severe contraction due to the combined effects of world recession and the collapse of its former markets in the Soviet Union.

Recovery is now underway. The economy is growing once more, although unemployment is high at 18 per cent. The new administration has declared its firm intention to reduce public spending and, in consequence, subsidies to farmers have been reduced.

THE REPUBLIC OF FINLAND	
STATUS	Republic
AREA	337,030 sq km (130,095 sq miles)
POPULATION	5,095,000
CAPITAL	Helsinki
LANGUAGE	Finnish, Swedish
RELIGION	87% Evangelical Lutheran, Eastern Orthodox minority
CURRENCY	markka (Finnmark) (FIM)
ORGANIZATIONS	Council of Europe, EEA, EU, OECD, UN

ÅLAND	
STATUS	Self-governing Island Province of Finland
AREA	1,505 sq km (581 sq miles)
POPULATION	24,231
CAPITAL	Mariehamn

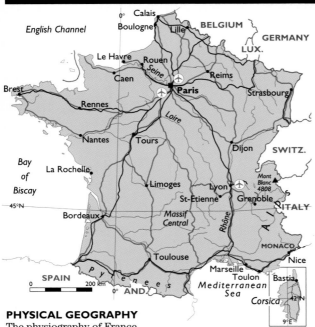

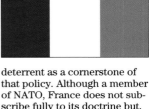

PHYSICAL GEOGRAPHY
The physiography of France shows considerable variation, ranging from the high peaks of the Alps and the Pyrenees to the flat plains of the north, the lowlands of the west coast, the granite moors of Brittany and the bleak Massif of central France.

CLIMATE
The climate of northern parts is temperate, but in the south is Mediterranean with warm dry summers and mild winters with some rainfall.

POLITICS AND RECENT HISTORY
In March 1993 the Socialist government was heavily defeated by a right wing alliance. Edouard Balladur, a protégé of the right wing leader Jacques Chirac, became prime minister. The association of socialist President Mitterrand, Head of State since 1981, and a Conservative prime minister was more successful than had been predicted. Although Balladur initially retained his popularity, despite the need for austerity measures, he lost ground in the run up to the 1995 presidential election. Both he and Chirac stood and, in the first round of voting, Chirac gained more votes that Balladur, although the Socialist candidate, Lionel Jospin, led. In the run off, Chirac comfortably defeated Jospin and succeeded Mitterrand as the fifth president of the fifth republic. Alain Juppé was appointed prime minister.

France continues to adopt a fiercely independent role in its international outlook, and maintains its own nuclear

THE FRENCH REPUBLIC	
STATUS	Republic
AREA	543,965 sq km (209,970 sq miles)
POPULATION	57,850,000
CAPITAL	Paris
LANGUAGE	French
RELIGION	90% Roman Catholic, Protestant, Muslim, Jewish minorities
CURRENCY	French franc (FRF)
ORGANIZATIONS	Council of Europe, EEA, EU, G7, NATO, OECD, UN, WEU

deterrent as a cornerstone of that policy. Although a member of NATO, France does not subscribe fully to its doctrine but, nevertheless, maintains powerful armed forces. The nation contributes more manpower than any other to the UN Protection Force in Bosnia.

ECONOMY
France is the fourth industrial power in the world after the USA, Japan and Germany. Its industries include iron and steel, chemicals, vehicles, aeronautics and armaments as well as food processing, electronics, luxury goods, fashion and perfumes. Energy has been provided by reserves of coal, oil and natural gas but in recent years other sources of energy have increased in importance, such as tidal power at the Rance estuary in Brittany, hydroelectric power in the mountains and nuclear energy using uranium from French mines.

For a long time France kept foreign investors at bay, but in recent years the policy has been reversed and France now equals Britain in popularity for direct foreign investment. The major reason for France now welcoming foreign firms is the need to find work for its unemployed. The attraction for foreign firms is France's geographically central position, its favourable corporate tax rate and the efficient telecommunications and transport network. The high-speed rail link between Paris and Lille was inaugurated in the summer of 1993 and the connection with London through the Channel Tunnel became effective in 1995. Foreign companies now account for 27 per cent of manufacturing output and 30 per cent of exports.

Despite its undoubted strength, the new French government faces some severe problems. The most pressing is unemployment which, at 12 per cent, is one of the highest among developed nations. Chirac has announced plans aimed at reducing the numbers of jobless, which include generous payments to employers who take on employees who have been unemployed for more than a year. At the same time, France has an uncomfortably large budget deficit and it is not easy to see how electoral promises, such as increasing public sector pay, can be ful-

filled. It may become necessary to curb some of the traditional protectionist policies involving subsidies to agriculture and industry. Pressing ahead with the ambitious privatization programme will assist this process.

Although France is highly industrialized, its national cultural identity is linked to agriculture. France is the world's second largest exporter of agricultural products, after the USA, and is the world's largest exporter of wines as well as being a major exporter of wheat, barley and sugar beet. Its tradition in agriculture gave rise to the world famous French cuisine. It is estimated that more than a quarter of the workforce derives an income from agricultural pursuits, whether they are from the huge sugar beet estates in the northeast, the small farms in Brittany or smallholdings in the Pyrenees. Over the years, however, there has been a steady drift of labour, mainly young people, from rural to urban areas. In 1990 there were a million farms in France compared with one and a half million in

1970. The trend has been matched with falls in sales of farming equipment – tractor sales have fallen by over 60 per cent in the past 10 years.

Although there is a growing disillusionment with the European Union among the French population, the government remains committed to eventual monetary union and, despite considerable difficulties, has kept the franc in the European exchange rate mechanism.

FRANCE – OVERSEAS TERRITORIES

WALLIS AND FUTUNA ISLANDS

STATUS	Self-governing Overseas Territory of France
AREA	274 sq km (106 sq miles)
POPULATION	14,100
CAPITAL	Mata-Utu

NEW CALEDONIA

STATUS	Overseas Territory of France
AREA	19,105 sq km (7,375 sq miles)
POPULATION	164,173
CAPITAL	Nouméa

FRENCH POLYNESIA

STATUS	Overseas Territory of France
AREA	3,940 sq km (1,520 sq miles)
POPULATION	188,814
CAPITAL	Papeete

ST PIERRE & MIQUELON

STATUS	Territorial Collectivity of France
AREA	241 sq km (93 sq miles)
POPULATION	6,392
CAPITAL	St Pierre

MAYOTTE

STATUS	Territorial Collectivity of France
AREA	376 sq km (145 sq miles)
POPULATION	85,000
CAPITAL	Dzaoudzi

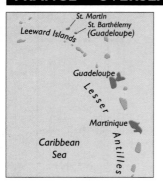

GUADELOUPE

STATUS .	**Overseas Department of France**
AREA	**1,780 sq km**
	(687 sq miles)
POPULATION	**400,000**
CAPITAL	**Basse-Terre**

FRENCH GUIANA

STATUS .	**Overseas Department of France**
AREA	**91,000 sq km**
	(35,125 sq miles)
POPULATION	**114,808**
CAPITAL	**Cayenne**

MARTINIQUE

STATUS .	**Overseas Department of France**
AREA	**1,079 sq km**
	(417 sq miles)
POPULATION	**359,572**
CAPITAL	**Fort-de-France**

RÉUNION

STATUS .	**Overseas Department of France**
AREA	**2,510 sq km**
	(969 sq miles)
POPULATION	**624,000**
CAPITAL	**Saint-Denis**

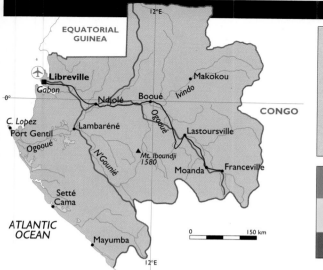

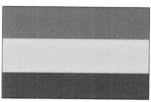

PHYSICAL GEOGRAPHY

Beyond the coastal strip, Gabon is a country of low plateaux, of which approximately 75 per cent is forest covered. Mt. Iboundji is the highest point, at 1,580 m (5,185 ft).

CLIMATE

The climate is hot and tropical with average annual temperatures of 25°C (77°F). Rainfall is heavy for most of the year but there is a pronounced dry season from June to August.

POLITICS AND RECENT HISTORY

President Omar Bongo, previously Albert-Bernard Bongo, came to power in 1967 and has retained the presidency since then. Until 1990 Gabon was a one-party state – under the Gabonese Democratic Party (PDG). Following widespread unrest during the 1980s and opposition to one-party rule, a new constitution was approved, allowing for multi-party elections. These were held in late 1990 and gave the PDG a narrow victory. Prime Minister Casimir Oye M'ba formed a government of national unity and included opposition representatives in his cabinet. In the December 1993 presidential elections, President Bongo regained the post with 51 per cent of the vote.

ECONOMY

Gabon is one of the most prosperous states in Africa, largely due to its valuable mineral resources and a relatively small population. The Gabonese economy is heavily dependent on oil, which has afforded the nation a measure of prosperity not shared by many African states. The country is the third largest oil producer in sub-Saharan Africa, after Nigeria and Angola. Nevertheless, reserves are not large and, although a new field at Avocette came into production in 1993, long-term prospects for the industry are poor. Gabon also exploits man-

THE GABONESE REPUBLIC	
STATUS	Republic
AREA	267,665 sq km (103,320 sq miles)
POPULATION	1,012,000
CAPITAL	Libreville
LANGUAGE	French, Bantu dialects, Fang
RELIGION	60% Roman Catholic
CURRENCY	CFA franc (C Africa) (XAF)
ORGANIZATIONS	OAU, OPEC, UN

ganese and uranium deposits, the former assisted by the opening of the Trans-Gabonais Railway in 1986, a major engineering project linking Libreville on the coast to Franceville in the interior. Road transport is primitive; there is no route linking Libreville and Port Gentil and transportation by river is common.

Agriculture contributes little to the economy, most farming is subsistence, and nearly half the population lives in towns. There are state-run plantations growing oil palms, bananas, sugar cane and rubber. Timber is an important export commodity, sold mostly to France, Gabon's major trading partner.

France supplies nearly half the country's total imports, and French influence is evident everywhere. Investment and expertise from France are particularly dominant in the mining and energy industries, and an 800-strong French military base is maintained in Libreville. In recent years, Gabon has sought links elsewhere. US investment is large, and relations are being strengthened with the Arab states and African neighbours.

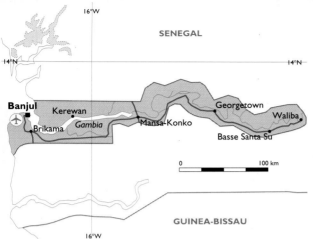

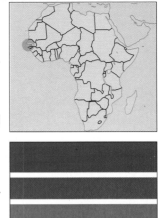

PHYSICAL GEOGRAPHY

The Gambia occupies a low-lying strip, 470 km (292 miles) long, of territory bordering the river of the same name. Nowhere is the territory wider than 50 km (30 miles). Long sandy beaches are backed by mangrove swamps along the river, beyond which is savannah and tropical woods.

CLIMATE

Two distinct seasons are evident: a warm dry season from November to mid-May and the rainy humid months from July to October, with rainfall of around 1,300 mm (51 inches). Temperatures average about 23–27°C (73–81°F) throughout the year.

POLITICS AND RECENT HISTORY

This former British colony became independent in 1965 and assumed the status of a republic, within the Commonwealth, in 1970. The head of state since independence was Sir Dawda Jawara, who was re-elected as president leading the People's Progressive Party in the general elections of May 1992.

Ten years previously, The Gambia had joined with Senegal in the Confederation of Senegambia. This short-lived experiment failed because of The Gambia's reluctance to move towards political union. Since then, The Gambia has sought to strengthen ties with Nigeria and increased its participation in the Economic Community of West African States.

The Gambia was regarded as one of Africa's most stable nations, but in July 1994 Sir Dawda Jawara was removed from power in a military coup led by Captain Yahya Jammeh. The military government has revealed massive corruption by the previous administration. It has appointed civilian ministers and announced a return to civilian rule in 1996.

THE REPUBLIC OF THE GAMBIA	
STATUS	Republic
AREA	10,690 sq km (4,125 sq miles)
POPULATION	1,026,000
CAPITAL	Banjul
LANGUAGE	English, Madinka, Fula, Wolof
RELIGION	90% Muslim, Christian and animist minorities
CURRENCY	dalasi (GMD)
ORGANIZATIONS	Comm., ECOWAS, OAU, UN

ECONOMY

The economy is based on groundnuts, which account for 80–90 per cent of exports and occupy 60 per cent of the cultivated land area. Three-quarters of the population rely particularly on groundnut production, together with subsistence farming and livestock rearing, for a livelihood. The dependence on revenue from groundnut sales means the economy is vulnerable to international price fluctuations as well as falls in output at times of drought.

In general there has been a failure to diversify from monoculture. Manufacturing accounts for less than 7 per cent of gross domestic product. The fisheries sector, however, is being improved and cotton and palms are grown.

Until the coup, tourism had provided the main area of economic growth – 4,500 tourists at a time visited the country. This fell drastically when the British government warned against travel to the country because of the instability. Export of vegetable produce was also affected because of the reduction in air traffic. This situation is, however, no more than a temporary setback.

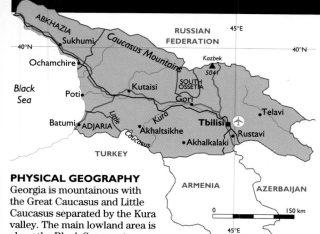

PHYSICAL GEOGRAPHY
Georgia is mountainous with the Great Caucasus and Little Caucasus separated by the Kura valley. The main lowland area is along the Black Sea coast.

CLIMATE
The climate is generally warm, except where modified by altitude, with most rain falling in western regions.

POLITICS AND RECENT HISTORY
In April 1991 the Georgian Socialist Republic declared its independence from the former Soviet Union. Eduard Shevardnaze, the former Soviet foreign minister, returned to Georgia in early 1992 when the elected president, Zviad Gamsakhurdia, was overthrown by paramilitary forces. Later, in October 1992, Shevardnaze was elected unopposed to the presidency.

The political scene has been dominated by ethnic rivalry: Georgians make up 70 per cent of the population, but there are significant minorities in South Ossetia, Abkhazia and Adjaria who seek autonomy. The most serious conflict has been in Abkhazia where, in 1993, rebels under the leadership of former president Gamsakhurdia gained control. Shevardnaze sought help from Russia and agreed to join the Commonwealth of Independent States. Russian forces, with the agreement of Western nations, entered Abkhazia to maintain the peace, following which Gamsakhurdia's death, probably by suicide, was announced. Stability is gradually returning and the enthusiasm of separatists appears to be waning.

ECONOMY
The Georgian economy has been badly damaged by strife and armed conflict. Economic output plummeted, the country ran a massive budget deficit, hyper-inflation reached alarming proportions and the coupon currency became virtually worthless. Many Georgians live in poverty, and the numbers of impoverished people have increased with the thousands of refugees who fled Abkhazia during the fighting.

Georgia lacks energy resources, although it has reserves of coal so far unexploited. The government, short of currency, has sought barter arrangements with neighbouring states, concluding arrangements with Azerbaijan to supply oil, and with Iran and Turkmenistan to supply gas.

The potential for prosperity remains based particularly on the agricultural produce of the rich plains along the Black Sea, where tea is a major crop.

As a fragile stability returns, there are some signs of recovery. Inflation, for example, is down to manageable proportions and the country follows International Monetary Fund and World Bank advice. A boost may come with the construction of an oil pipe-line from Azerbaijan, across Georgia, to the Black Sea port of Batumi. The Russians, however, want the pipe-line to cross their territory.

THE REPUBLIC OF GEORGIA	
STATUS	Republic
AREA	69,700 sq km (26,905 sq miles)
POPULATION	5,471,000
CAPITAL	Tbilisi
LANGUAGE	70% Georgian, 8% Armenian, 6% Russian, 6% Azeri
RELIGION	Orthodox Christian
CURRENCY	coupon
ORGANIZATIONS	CIS, UN

ABKHAZIA	
STATUS	Autonomous Republic of Georgia
AREA	8,600 sq km (3,320 sq miles)
POPULATION	538,000
CAPITAL	Sukhumi

ADJARIA	
STATUS	Autonomous Republic of Georgia
AREA	3,000 sq km (1,160 sq miles)
POPULATION	382,000
CAPITAL	Batumi

PHYSICAL GEOGRAPHY

The North German Plain, extending across the breadth of Germany, is a mixture of fertile farmland and sandy heaths. The uplands of the south include the Black Forest, the Swabian and the Bavarian Alps.

CLIMATE

The climate is temperate, with continental tendencies in eastern parts, where winters are colder. Rainfall is spread evenly throughout the year.

POLITICS AND RECENT HISTORY

One of the most dramatic political events of recent years was the reunification of East and West Germany. During 1989 increasing public unrest in East Germany led, in October, to the downfall of the long-serving Communist leader Erich Honecker and, a month later, to the destruction of the Berlin Wall. Reunification required the agreement of the World War II occupying powers. The former Soviet Union, politically weakened itself, capitulated and agreed to German reunification within the North Atlantic Treaty Organization and undertook to withdraw all Warsaw Pact troops from German soil. Monetary union was achieved in July 1990 and political union completed in October of the same year.

Chancellor Helmut Kohl

THE FEDERAL REPUBLIC OF GERMANY	
STATUS	Federal Republic
AREA	356,840 sq km (137,740 sq miles)
POPULATION	81,373,000
CAPITAL	Berlin (seat of government Berlin/Bonn)
LANGUAGE	German
RELIGION	45% Protestant, 40% Roman Catholic
CURRENCY	Deutsche mark (DM)
ORGANIZATIONS	Council of Europe, EEA, EU, G7, NATO, OECD, UN, WEU

became the first leader of the newly united Germany at elections in December 1990. His government lost popularity because of the harsh measures needed to achieve integration. Westerners resented the belt-tightening demanded of them, while in the east the population complained of the slow progress towards equality. Despite this, Chancellor Kohl managed to hold on to power by the narrowest of margins in the 1994 elections. He again leads a coalition of his own Christian Democratic Union, the Christian Social Union and the Free Democratic Party.

On a wider front Germany, whose constitution prevents its active participation in United Nations (UN) operations, has signified a wish to end this status and to participate more fully on the world stage. In part this is motivated by its ambition to gain a permanent seat on the UN Security Council.

ECONOMY

The economic price of unification was grievously underestimated. West Germany was the economic power house of western Europe and with its gross domestic product averaging over DM 35,000 per head could, it was thought, comfortably absorb and modernize the East German economy.

The reality has been different, and not merely because of world recession. East German industry, although probably the most successful among the eastern states, was nevertheless grossly inefficient by western standards. It was overmanned, needed new investment, and many of its traditional markets were lost with the break-up of the communist trading bloc.

In the years immediately following unification, the German economy was strained by the burden. Unemployment rose to 7 per cent in the west and to 15 per cent in the east. Economic growth shrunk due to recession in Europe generally, where adjustments to other currencies left German goods relatively more expensive. The situation was aggravated by large payments to Russia, to help meet the costs of the removal of its troops from east Germany.

Throughout this period, the government and Bundesbank kept tight control of fiscal and monetary policies. Interest rates have remained relatively high to protect the mark and inflation has been kept under control.

It is clear that the worst of the economic problems of reunification are now in the past. The east German economy, which turned round in 1993, is growing fast and the massive programme of privatization of east German state-run businesses was virtually complete by the end of 1994. Even so, the disparity between East and West Germany was so great that it will be many years before true economic unity is achieved.

BADEN-WÜRTTEMBERG

STATUS	State (Land)
AREA	35,730 sq km (13,790 sq miles)
POPULATION	9,400,000
CAPITAL	Stuttgart

BAYERN (BAVARIA)

STATUS	State (Land)
AREA	70,545 sq km (27,230 sq miles)
POPULATION	11,000,000
CAPITAL	Munich

BERLIN

STATUS	State (Land)
AREA	883 sq km (341 sq miles)
POPULATION	3,400,000
CAPITAL	Berlin

BRANDENBURG

STATUS	State (Land)
AREA	29,059 sq km (11,220 sq miles)
POPULATION	2,700,000
CAPITAL	Potsdam

BREMEN

STATUS	State (Land) City Territory
AREA	404 sq km (156 sq miles)
POPULATION	700,000
CAPITAL	Bremen

HAMBURG

STATUS	State (Land) City Territory
AREA	755 sq km (291 sq miles)
POPULATION	1,600,000
CAPITAL	Hamburg

HESSEN (HESSE)

STATUS	State (Land)
AREA	21,115 sq km (8,150 sq miles)
POPULATION	5,600,000
CAPITAL	Wiesbaden

MECKLENBURG-VORPOMMERN

STATUS	State (Land)
AREA	23,838 sq km (9,204 sq miles)
POPULATION	2,100,000
CAPITAL	Schwerin

NIEDERSACHSEN (LOWER SAXONY)

STATUS	State (Land)
AREA	47,425 sq km (18,305 sq miles)
POPULATION	7,200,000
CAPITAL	Hanover

NORDRHEIN-WESTFALEN

STATUS	State (Land)
AREA	34,070 sq km (13,150 sq miles)
POPULATION	16,900,000
CAPITAL	Dusseldorf

RHEINLAND-PFALZ

STATUS	State (Land)
AREA	19,840 sq km (7,660 sq miles)
POPULATION	3,700,000
CAPITAL	Mainz

SAARLAND

STATUS	State (Land)
AREA	2,575 sq km (994 sq miles)
POPULATION	1,100,000
CAPITAL	Saarbrucken

SACHSEN (SAXONY)

STATUS	State (Land)
AREA	18,337 sq km (7,080 sq miles)
POPULATION	4,900,000
CAPITAL	Dresden

SACHSEN-ANHALT

STATUS	State (Land)
AREA	20,445 sq km (7,894 sq miles)
POPULATION	3,000,000
CAPITAL	Halle

SCHLESWIG-HOLSTEIN

STATUS	State (Land)
AREA	15,710 sq km (6,065 sq miles)
POPULATION	2,600,000
CAPITAL	Kiel

THÜRINGEN (THURINGIA)

STATUS	State (Land)
AREA	16,251 sq km (6,275 sq miles)
POPULATION	2,500,000
CAPITAL	Erfurt

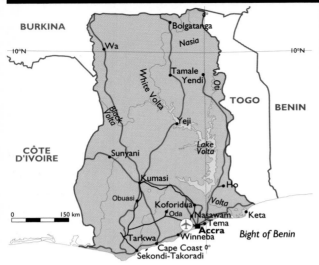

PHYSICAL GEOGRAPHY
Western Ghana is covered by dense rainforest. The terrain becomes more hilly to the north, culminating in a plateau averaging some 500 m (1,600 ft).

CLIMATE
The climate is tropical with temperatures in the range 21–32°C (70–90°F). The average rainfall is 2,000 mm (80 inches) on the coast, less inland. The coastal climate has no real seasonal variation.

POLITICS AND RECENT HISTORY
In 1957 Ghana, once known as the Gold Coast, became the first African state to gain full independence from Britain. Dr Kwame Nkrumah was the first ruler; accused of corruption, he was overthrown by a military coup in 1966. Since then politics have been dominated by military governments, although in recent years there has been a trend towards political pluralism. Jerry John Rawlings is currently head of state, having come to power in 1981. In 1992 a new constitution, approved by referendum, introduced a form of government based on the US system. Rawlings kept the presidency in elections, nominally multi-party but largely boycotted.

ECONOMY
Cocoa is the principal cash crop. Ghana was the world's leading producer until it was overtaken by the Côte d'Ivoire in the late 1970s. Over the last decade low world prices for cocoa have affected the economy, although prices seemed set to improve.

Gold has been mined in Ghana for 1,000 years. After falling production in the early 1980s, output is now booming and is approaching unprecedented figures of two million troy ounces per annum. New discoveries in the Ashanti fields assure gold wealth for many years to come. The commodity now outstrips cocoa as the leading export earner. The Ashanti Goldfields Corporation, the leading operator in Ghana, has recently been privatized.

Timber is a major source of earnings, and the government is set to spend $16.8 million on forest resource management. Other products include manganese, revived to meet the demand in battery production. Aluminium is smelted using hydro-electric power generated at the Akosombo Dam which, when built, formed Lake Volta, one of the world's largest artificially created lakes. Offshore oil has yet to be economically developed.

The Ghanaian population is mostly involved in subsistence agriculture. The people endure a devalued currency and inflation of 25 per cent, as the government tries to meet the advice of the World Bank and the International Monetary Fund.

THE REPUBLIC OF GHANA	
STATUS	Republic
AREA	238,305 sq km (91,985 sq miles)
POPULATION	16,446,000
CAPITAL	Accra
LANGUAGE	English, tribal languages
RELIGION	42% Christian
CURRENCY	cedi (GHC)
ORGANIZATIONS	Comm., ECOWAS, OAU, UN

PHYSICAL GEOGRAPHY

Greece is mountainous and over one-fifth of its area is comprised of numerous islands, 154 of which are inhabited.

CLIMATE

The summers are hot and dry. Winters are mild and wet, although the mountains experience heavy snowfalls.

POLITICS AND RECENT HISTORY

A military coup in 1967 brought an end to the Greek monarchy with the exile of King Constantine II. Seven years of dictatorship followed until, in 1974, a civilian government was restored.

For 30 years politics have been dominated by two rival leaders: Andreas Papandreou of the socialist Pasok party and Constantine Mitsotakis of the conservative New Democracy. The Mitsotakis government was forced to hold a general election in October 1993, after losing a working majority in parliament. In the election Andreas Papandreou, who had been ousted from office in 1989 following a series of scandals, returned to office after the Pasok party won a larger than expected majority.

In 1994 Papandreou ensured the survival of the government for its full four-year term by reaching agreement with a splinter group, Political Spring, on the election of Costis Stephanopoulos as president.

Greece has incurred the disapproval of its European Union (EU) partners over its attitude to the Former Yugoslav Republic of Macedonia, but is now softening its approach. It appears ready to lift the blockade, may not press its objection over the name Macedonia, and may start a dialogue with the republic.

ECONOMY

THE HELLENIC REPUBLIC	
STATUS	Republic
AREA	131,985 sq km (50,945 sq miles)
POPULATION	10,350,000
CAPITAL	Athens (Athina)
LANGUAGE	Greek
RELIGION	97% Greek Orthodox
CURRENCY	drachma (GRD)
ORGANIZATIONS	Council of Europe, EU, EEA, NATO, OECD, UN

National income is derived primarily from tourism and services. Other important sources are agriculture, remittances from Greek emigrants and a large merchant shipping fleet.

Agricultural exports include fruit, vegetables, cotton, tobacco, olives, wine and cheese. The food processing and fish farming industries are growth areas. However, many farms and businesses are traditionally small family concerns, vulnerable to rising costs and competition, and exports remain low compared with imports.

The war in neighbouring ex-Yugoslavia, with the United Nations' embargo on trade to Serbia, has lost Greece an important market. The closed frontier has also made it necessary to use new routes into Europe which are longer and more costly.

Overall, the Greek economy is not flourishing. It suffers from slow growth, high inflation and a substantial fiscal deficit. EU assistance has been provided both directly and as development aid for road improvement, the construction of underground railways and a new airport in Athens.

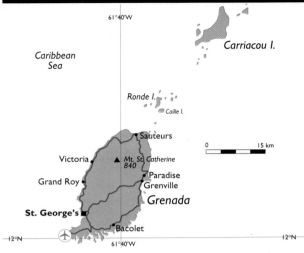

PHYSICAL GEOGRAPHY

Grenada, whose territory includes the Southern Grenadines, is a volcanic island of wooded hills and fast-flowing streams.

CLIMATE

The climate is warm throughout the year, with average temperatures above 25°C (77°F). Rainfall is plentiful, mostly falling in summer.

GOVERNMENT

Formerly a British colony, Grenada became self-governing in 1967 and achieved full independence within the Commonwealth in 1974. The government at the time was led by Prime Minister Sir Eric Gairy, whose regime became more authoritarian and was overthrown in a coup led by Maurice Bishop and the New Jewel Movement. However, Bishop's revolutionary government caused even more concern, particularly to the neighbouring Caribbean islands and the USA, which foresaw Grenada becoming a Marxist state on the Cuban model. Events came to a head in 1983 when disagreements between Bishop, other Marxists and the army prompted a military coup and Bishop was killed. At the

request of the Organization of Eastern Caribbean States (OECS), a US-led force carried out an invasion of the island lasting three days, restoring democracy in the form of an interim government. Parliamentary elections were held in 1984.

Since then political parties have split and realigned. Nicholas Braithwaite of the National Democratic Congress (NDC), as leader of the largest single party, was able to form a coalition government with the support of the National Party. He resigned as prime minister in February 1995 and was replaced by George Brizan, who become leader of the NDC in September 1994. The June 1995 general election was won by the opposition New National Party, defeating the

NDC and the United Labour Party.

Grenada is one of the four countries of the OECS in favour of moves towards political and economic integration in the Windward Islands.

ECONOMY

The economy is firmly based on agriculture. The most important crop is nutmeg, and Grenada and Indonesia are the world's only suppliers. Other major crops are citrus fruits, bananas and cocoa. Tourism has increased and is now a potentially important revenue earner, although the latest figures indicate a poor record of profitability. Some light industries, such as furniture, garment and soft drinks production, have been developed.

In 1994, Grenada began producing nutmeg oil, making use of defective nuts which were previously destroyed. It is hoped this new development will boost the nutmeg industry which has suffered from price falls in the international market for some years.

Although inflation rates are low, at less than 5 per cent, a quarter of the workforce is unemployed, and the indications are that this figure will worsen.

GRENADA	
STATUS	Commonwealth State
AREA	378 sq km (146 sq miles)
POPULATION	92,000
CAPITAL	St George's
LANGUAGE	English, French patois
RELIGION	Roman Catholic majority
CURRENCY	E Caribbean dollar (XCD)
ORGANIZATIONS	Caricom, Comm., OAS, UN

PHYSICAL GEOGRAPHY

Northern parts of Guatemala are lowland tropical forests. To the south lie high mountain ranges with volcanic peaks, some of which are active. A coastal plain borders the Pacific.

CLIMATE

The northern lowlands and the Pacific coastal strip have a hot tropical climate, but the central highlands are cooler with average daily temperatures of around 20°C (68°F). Rainfall is heaviest in summer.

POLITICS AND RECENT HISTORY

Military dictatorship gave way to civilian rule in 1985, but accusations of corruption within government have not diminished and Guatemala is said to have one of the world's worst records in human rights.

In May 1993 President Jorge Serrano dissolved Congress and attempted to rule as a dictator, expecting support from the army. Instead, a broad coalition of the military, businessmen and trade-unionists compelled him to stand down. Congress elected Ramiro De León Carpio to serve the two-and-a-half years remaining of Serrano's term. President De León, a human rights activist, initially enjoyed widespread support but, without political backing, he has found it difficult to achieve his objective of ridding Congress of its corrupt representatives.

Although civil war between the government and the rebel Guatemalan National Revolutionary Unity (URNG) continues, United Nations sponsored peace talks in Mexico are beginning to progress, albeit slowly. One of the sticking points is the status of the majority indigenous Indian population, who are still suffering human rights violations. It seems likely that the URNG, although not fielding candidates themselves, will support a left wing candidate at the next presidential elections. This is the result of consternation at the likely candidature of Efrain Rios Montt, the right wing president of Congress and a former military dictator during the worst years of government atrocities.

ECONOMY

The Guatemalan economy is founded primarily on agriculture, but there is also a substantial manufacturing industry which includes textiles, paper and pharmaceuticals. The leading crop is coffee but bananas, cotton and sugar are also important. In recent years there has been a growth in the export of high-value fresh fruit and vegetables, mainly to the USA.

Amid controversy, the government is seeking to develop the tropical forests of Petén and to encourage settlement there. Environmentalists fear the destruction of large tracts of virgin forest. Meanwhile illegal logging operations and drug trafficking plague this region.

Guatemala suffers from widespread poverty and requires support from the International Monetary Fund, who in turn have demanded tax reforms.

THE REPUBLIC OF GUATEMALA	
STATUS	Republic
AREA	108,890 sq km (42,030 sq miles)
POPULATION	10,322,000
CAPITAL	Guatemala City
LANGUAGE	Spanish, Indian languages
RELIGION	75% Roman Catholic, 25% Protestant
CURRENCY	quetzal (GTQ)
ORGANIZATIONS	CACM, OAS, UN

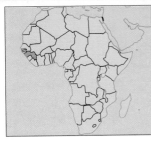

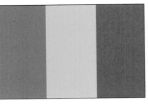

PHYSICAL GEOGRAPHY
Mangrove swamps are found along the coastal plains. The Guinea Highlands dominate the south, mountains and plateaux the west. In the east, savannah plains are drained by the upper Niger river system.

CLIMATE
The climate is tropical, the coastal zones experiencing heavy rainfall of around 4,369 mm (172 inches) annually, falling mainly between June and October. High temperatures of around 24–32°C (75–90°F) occur throughout the year.

POLITICS AND RECENT HISTORY
Guinea, French Guinea until its independence in 1958, was ruled for 26 years under a repressive regime, headed by President Ahmed Sékou Touré until his death in 1984. A military coup followed, led by Colonel Lansana Conté who became president, and links with the west were opened after years of isolation. Until 1992, political control was through a Military Committee. At that time, President Conté yielded power to a Transitional Committee for National Recovery, a temporary government designed to promote democratic elections. At the end of 1993, the first-ever multi-party presidential elections were held and Conté regained the presidency. In early 1995, President Conté assumed the rank of general, having resigned from the army in 1993, in order to stand in the presidential elections.

ECONOMY
On gaining power in 1984, Conté embarked on a number of economic reforms and sought to expand economic links with the European Union. He was supported by the International Monetary Fund and the World Bank, but Guinea remains one of the world's poorest countries. Corruption is still endemic in the civil service. Privatization of industries has failed through inexperienced management and industrial production has consequently ceased.

Two-thirds of Guinea's population are occupied in agriculture, chiefly subsistence farming, with bananas and pineapples among the main cash products. The agricultural sector has declined in the last decade due to mismanagement of investment. In 1984 the country was self-sufficient in basic foodstuffs, but it now has to import rice, the staple food.

The remaining industrial sector is mining. Guinea is rich in natural resources and possesses the world's largest reserves of bauxite, which accounts for 80 per cent of export revenue. The government is promoting the development of a bauxite mining and alumina smelting project near the world's biggest open-pit bauxite mine at Boké, but costly rail and harbour improvements will be required. Diamonds are also important and mining has increased in recent years in response to demand. Gold mining began in 1988. A new open-pit gold mine began production in 1995 and a substantial new mine is expected to open in 1996. Extensive high-ground iron ore deposits in the south await exploitation.

THE REPUBLIC OF GUINEA	
STATUS	Republic
AREA	245,855 sq km (94,900 sq miles)
POPULATION	6,306,000
CAPITAL	Conakry
LANGUAGE	French, Susu, Manika
RELIGION	85% Muslim, 10% animist, 5% Roman Catholic
CURRENCY	Guinea franc (GNF)
ORGANIZATIONS	ECOWAS, OAU, UN

PHYSICAL GEOGRAPHY

Guinea-Bissau encompasses the Bijagós archipelago. The terrain is generally low-lying, with numerous estuaries and swamps. The plains inland are thickly forested while to the east there are savannah plateaux.

CLIMATE

The climate is tropical. The dry season, when the Harmattan wind blows, is from December to April. The wet season, with high humidity, is from May to November. Annual rainfall is around 2,000–3,000 mm (79–118 inches).

POLITICS AND RECENT HISTORY

The independence of Guinea-Bissau, formerly Portuguese Guinea, was recognized in 1974. Seventeen years of one-party rule by the *Partido Africano da Independencia da Guiné e Cabo Verde* (PAIGC) ended in 1991 when opposition parties were legalized, and in 1993 the constitution was revised to permit multi-party elections. These took place in July–August 1994, with a decisive victory for the PAIGC and a return to power for President João Bernardo Vieira, who has held office since 1978.

ECONOMY

The economy is largely one of subsistence farming, which involves over 80 per cent of the active population. The main cash crops are groundnuts, palm kernels and cashews. Many industries have yet to be developed; there are untapped reserves of phosphates and bauxite, and also of offshore oil. Fishing and tobacco industries are being encouraged, and the Portuguese are contributing to a new forestry scheme.

Guinea-Bissau relies heavily on foreign aid, with one of the highest per capita rates in the world. The UK has given financial assistance towards helping refugees from Casamance in Senegal. The government has not paid off arrears to the World Bank and is in a state of virtual bankruptcy. Industrial unrest in recent years has paralysed the country, as the five main trade unions negotiated with ministers. Civil servants and government employees are no longer paid regularly, and teachers have held strikes for an improvement to their meagre wages.

Guinea-Bissau has formally applied to join the Franc Zone, but because of chronic public financial problems it is probably no closer to joining than when it first applied in 1985. The average annual inflation rates of other Franc Zone countries are 5 per cent or less, but in Guinea-Bissau inflation in the early 1990s has averaged 110 per cent.

THE REPUBLIC OF GUINEA-BISSAU	
STATUS	Republic
AREA	36,125 sq km (13,945 sq miles)
POPULATION	1,028,000
CAPITAL	Bissau
LANGUAGE	Portuguese, Guinean, Creole
RELIGION	Muslim and animist majority, Roman Catholic minority
CURRENCY	Guinea-Bissau peso (GWP)
ORGANIZATIONS	ECOWAS, OAU, UN

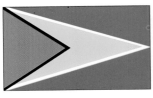

PHYSICAL GEOGRAPHY

Beyond the lowland coastal belt of Guyana, the land rises inland to high savannah uplands and forested mountains.

CLIMATE

The climate is tropical with hot, wet and humid conditions, which are modified along the coast by sea breezes.

POLITICS AND RECENT HISTORY

Guyana, formerly British Guiana, gained independence from Britain in 1966 and became a republic in 1970.

Elections have not been held with the regularity that the constitution demands, and those due in December 1991 were postponed until October 1992. The victorious candidate for the presidency was Dr Cheddi Jagan, of the left wing People's Progressive Party, who defeated the previous incumbent Desmond Hoyte of the People's National Congress. Despite his Marxist background, Dr Jagan has not reversed his predecessor's policies and has declared his faith in democracy and pluralism. Even so, he has been the subject of criticism in the Guyana press, especially over delays in

decisions about the privatization of the Guyana Electricity Corporation, and his predecessor has made allegations of favouritism towards the Indian sector of the population. Racial rivalry between people of African descent and those of Indian origin has always been a dominant issue in Guyana's politics.

ECONOMY

Guyana's economic wealth is concentrated in the coastal areas. Agriculture, in which sugar and rice predominate, is the basis of the economy. Sugar production has generally declined and although there has been a modest recovery in recent years, this is not expected to be sustained in the face of stagnant world prices.

However, rice production has increased, with exports doubling in the early 1990s. Guyana also enjoys considerable mineral wealth and although bauxite exports, the principal commodity, have declined in value, this has been offset by increased exports of gold. Gold production soared in 1993 and 1994, as a result of increased output at the new Omai mine.

Generally, the Guyanan economy has grown since 1990. Continued prosperity is mostly dependent on the encouragement of inward investment, initiated by the previous administration, and the programme of privatization, approved by parliament in July 1993. However, plans to sell the government's shares in state enterprises such as the Demerara Distillers, Guyana Airways and in particular Guysuco (the Guyana Sugar Company) could spark some controversy.

THE CO-OPERATIVE REPUBLIC OF GUYANA	
STATUS	Co-operative Republic
AREA	214,970 sq km (82,980 sq miles)
POPULATION	816,000
CAPITAL	Georgetown
LANGUAGE	English, Hindi, Urdu, Amerindian dialects
RELIGION	Christian majority, Muslim and Hindu minorities
CURRENCY	Guyana dollar (GYD)
ORGANIZATIONS	Caricom, Comm., UN

PHYSICAL GEOGRAPHY

Haiti, occupying western Hispaniola, is a mountainous and forested country. Its highest part, reaching over 2,600 m (8,500 ft), is near the Dominican border.

CLIMATE

Temperatures are consistently high in coastal areas but become cooler with altitude. Annual rainfall averaging 1,400 mm (55 inches) occurs throughout the year, with occasional thunderstorms.

POLITICS AND RECENT HISTORY

For three decades, until 1986, Haiti was ruled by the dictatorships of first Papa Doc François Duvalier and then his son Baby Doc Jean-Claude Duvalier, assisted by the murderous secret police militia. After brief military and interim governments, Haiti's first multiparty elections were held in December 1990 and Catholic priest Father Jean-Bertrand Aristide was elected President.

Less than a year later Aristide was ousted in a military coup led by General Raoul Cedras and police chief Colonel Michel François, who became the *de facto* rulers of

Haiti. International objection took the form of trade embargoes, but these caused huge deprivation to an already impoverished nation. By spring 1992 thousands of Haitians were fleeing the country by boat, seeking refuge in the USA from starvation and the regime of terror inflicted by François's secret police attachés. The US intervened, supporting the exiled Father Aristide, and a UN and arms embargo in June 1994 forced the military leaders to negotiate. Negotiations, however, were tortuous and only after diplomacy from the former US president Jimmy Carter, and occupation by US troops, did General Cedras and his associates agree to give up power and leave the country. President Aristide returned in October 1994, appointed

Smarck Michel as prime minister and resigned the priesthood in order to concentrate on his presidential duties. A UN force took over from the US contingent in 1995, with a mandate to preserve law and order while democracy is restored.

ECONOMY

More than three-quarters of Haiti's population fall within the World Bank's category of absolute poverty, worsened by the trade embargoes and withholding of foreign aid. Now that the political problems are resolved, President Aristide and his government are seeking to rebuild the devastated economy. The United States and France have provided financial assistance and agreements have been reached with the World Bank and International Monetary Fund. Foreign investment is being encouraged.

The economy is essentially agricultural, with income mainly derived from the export of coffee, sugar, cotton and cocoa. Before the political upheaval, the assembly of goods for the American market had grown in importance and the government will seek to re-establish this industry.

THE REPUBLIC OF HAITI	
STATUS	Republic
AREA	27,750 sq km (10,710 sq miles)
POPULATION	7,041,000
CAPITAL	Port-au-Prince
LANGUAGE	90% Creole, French
RELIGION	80% Roman Catholic, some Voodoo folk religion
CURRENCY	gourde (HTG)
ORGANIZATIONS	OAS, UN

PHYSICAL GEOGRAPHY
The terrain is rugged and mountainous, with peaks rising to over 2,500 m (8,200 ft). The lowland areas are along the Caribbean and Pacific coasts.

CLIMATE
Coastal regions are hot and humid, with the heaviest rainfall in the summer months, averaging 2,500 mm (98 inches) a year. The interior enjoys a cooler, drier and more temperate climate.

POLITICS AND GOVERNMENT
Honduras emerged from military dictatorship in 1980, and since then relatively stable civil rule has been maintained. Politically, the country is dominated by the *Partido Nacional* (PN) and the *Partido Liberal* (PL). In the 1989 elections, the PN were victorious and their candidate Rafael Callejas was appointed president.

Throughout the 1980s Honduran politics were affected by the civil war in neighbouring Nicaragua. Aid for the Contras passed through the country, in return for which Honduras received substantial US aid. However, the Honduran government was also a prime mover in achieving peace in Nicaragua.

The military exerts a powerful influence but its authority seems likely to wane following a landmark case in 1993 when military officers were, for the first time, tried in the civil courts for murder.

Elections took place in November 1993 and the PL's Carlos Reina ousted the PN in a landslide victory.

In early 1995, President Reina announced the formation of the country's first civilian police force, ostensibly to replace the military secret police who were disbanded in mid-1994, following accusations of human rights violations.

ECONOMY
Honduras is poor, but the former president, Callejas, has set the nation on the road towards growth, expansion of exports and low inflation. On the other hand, the country suffers massive unemployment, officially measured at 45 per cent.

The economy is essentially agricultural, and although Honduras has thrown off its banana republic image, that crop remains important, although less so than coffee. Overall, agriculture provides 70 per cent of export revenue, in which coffee and bananas predominate. Extensive storm damage to northern coastal areas in September 1993 adversely affected banana yields.

There are few natural resources although lead, silver and zinc are exported and exploitation of these resources, together with oil, may lead to a change in the traditional agriculture-based economy.

The country is dependent upon support from the International Monetary Fund (IMF) and the World Bank. US aid, which reached $40 million during the 1980s, has declined to a mere $2 million. The IMF and World Bank are encouraging a better-managed timber industry and diversification in agriculture. Tomato production has doubled, and a potentially profitable shrimp industry is growing. In further reforms, the government has announced the privatization of the state-owned electric company.

THE REPUBLIC OF HONDURAS	
STATUS	Republic
AREA	112,085 sq km (43,265 sq miles)
POPULATION	5,770,000
CAPITAL	Tegucigalpa
LANGUAGE	Spanish, Indian dialects
RELIGION	Roman Catholic majority
CURRENCY	lempira (HNL)
ORGANIZATIONS	CACM, OAS, UN

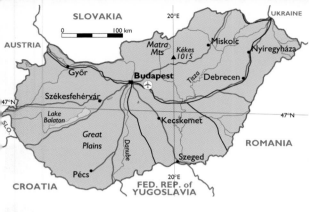

PHYSICAL GEOGRAPHY
West of the Danube, Hungary consists of rolling countryside. To the southeast lies the fertile Great Plain, while the highest terrain is in the northeast.

CLIMATE
The climatic regime is distinctly continental, with warm summers and cold winters. Rainfall is fairly evenly distributed throughout the year.

POLITICS AND RECENT HISTORY
Hungary's emergence from Communism began earlier than its east European neighbours. Janos Kadar, who had held power as Communist Party secretary since the Hungarian Uprising in 1956, was removed from office in 1988. The first free elections followed in 1990. The right wing Democratic Forum was victorious and ruled until 1994, in coalition with smaller parties. However, its popularity waned and in the 1994 election the Socialist Party, consisting of former Communists, gained an overall majority. Aware of national and international unease at power being concentrated in the hands of former Communists, the Socialists, under prime minister Gyula Horn, formed a coalition government with the Liberals.

ECONOMY
Economic reforms in the late 1980s gave Hungary a lead among the former Communist nations in the transformation to a market economy, with privatization based on substantial investment from western Europe. However, progress has faltered and the government has been forced to woo its own population with incentives for share ownership and a voucher system. The Hungarian economy has declined, with consistent falls in gross domestic product, inflation of over 20 per cent, unemployment of 13 per cent and a substantial trade deficit. Loans from the International Monetary Fund and the World Bank have been made on condition that the government institutes stringent financial policies.

A particular economic weakness is the decline in the agricultural sector, the basis of the Hungarian economy. This decline is associated with the difficult transfer to private ownership and the lack of capital to support the industry.

General economic decline has reached such proportions that the Socialist government has been forced to introduce an austere reform package, aimed at cutting government spending, reducing imports and boosting exports. These measures are bound to cause resentment among the population, as they will affect the generous welfare payments to which Hungarians have become accustomed. Even so, they are seen as essential if economic well-being is to be restored, and Hungary's desire to join the European Union is to be realised.

THE REPUBLIC OF HUNGARY	
STATUS	Republic
AREA	93,030 sq km (35,910 sq miles)
POPULATION	10,257,000
CAPITAL	Budapest
LANGUAGE	Hungarian (Magyar)
RELIGION	60% Roman Catholic, 20% Hungarian Reformed Church, Lutheran and Orthodox minorities
CURRENCY	forint (HUF)
ORGANIZATIONS	Council of Europe, UN

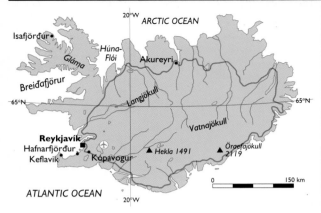

PHYSICAL GEOGRAPHY
Iceland is the northernmost island in Europe. One-tenth of the country is covered by ice caps, but there is also active volcanicity in the form of volcanoes, geysers and hot springs. Over half the population lives in the Reykjavík area.

CLIMATE
Although Iceland is moist and cool, with average summer temperatures of 9–10°C (48–50°F), it is protected from the harsh conditions normally associated with a northerly latitude by the Gulf Stream and prevailing southwesterly winds.

POLITICS AND RECENT HISTORY
Once under Danish sovereignty, Iceland was proclaimed a republic in 1944 and is now a parliamentary democracy with a president as head of state. In more than 30 years no single party has secured a majority in parliament (the *Althing*) and coalitions have become the rule.

In recent elections, the Independence Party, under Prime Minister David Oddson, has gained most seats, but in the 1995 elections its former partner, the Social Democrat Party, lost ground and the government is now a coalition of the Independence and centrist Progressive Party. Iceland is unlikely, in the short term, to seek membership of the European Union (EU), because the EU common fisheries policy is not seen as compatible with Iceland's interests. There is, however, growing support for membership within the Social Democrat party and the business community.

ECONOMY
The traditional mainstay of the Icelandic economy has been fishing, particularly cod, and the country is, therefore, subject to all the problems associated with heavy dependence on one commodity. Iceland jealously protects its fishing limits and imposes quotas on its own fishermen in order to protect stocks. This has resulted in a decline in the catch and the general standard of living, exports to its principal customer, the UK, having declined.

THE REPUBLIC OF ICELAND	
STATUS	Republic
AREA	102,820 sq km (39,690 sq miles)
POPULATION	263,000
CAPITAL	Reykjavík
LANGUAGE	Icelandic
RELIGION	93% Evangelical Lutheran
CURRENCY	Icelandic krona
ORGANIZATIONS	Council of Europe, EEA, EFTA, NATO, OECD, UN

However, there are signs of a general improvement. Although the national budget is in deficit, the economy is growing. Iceland's attempts to exploit fish stocks beyond its own waters have not met with success and it has failed to secure any fishing rights around Svalbard, from Norway. Iceland was once a great whaling nation and although it has resigned from the International Whaling Commission, it is unlikely to resume whaling because of the international disapproval that would be incurred.

Less than 2 per cent of the land is suitable for cultivation; the rest is used as grazing for livestock, Iceland being self-sufficient in meat and dairy products. Wool and sheepskin are exported, and there has been a gradual increase in light industry, such as the manufacture of blankets and knitwear. An aluminium smelting industry, based on imported bauxite, is run on hydro-electric power. Income is also derived from a growing tourist industry.

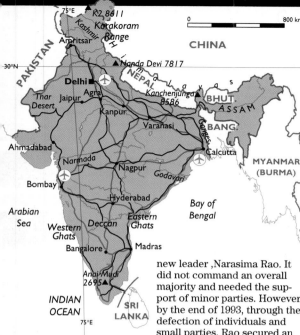

PHYSICAL GEOGRAPHY

The heart of the Indian peninsula is the Deccan plateau, bordered on either side by the Eastern and Western Ghats. The Ganges Valley dominates the northern lowlands, beyond which rise the Himalayas, the world's highest mountain range, with many peaks exceeding 6,000 m (19,685 ft).

CLIMATE

Within India there is great diversity of climatic conditions. Generally, the climate is tropical with monsoon rains in summer. In the pre-monsoon season, the heat becomes intense, with average temperatures in New Delhi reaching 38°C (100°F).

POLITICS AND RECENT HISTORY

Although the Congress Party suffered the assassination of its leader Rajiv Gandhi just before the last general election in June 1991, the party was able to keep its hold on power under a new leader, Narasima Rao. It did not command an overall majority and needed the support of minor parties. However, by the end of 1993, through the defection of individuals and small parties, Rao secured an absolute majority.

Rao's administration has been turbulent. He has been accused of financial impropriety and now has to contend with splits in the Congress party. Support for the Hindu Nationalist party has grown and the party gained control of Maharashtra province in the 1995 provincial elections, when Congress also lost control of Gujarat.

India has improved its relations with China and has reached agreement on the removal of troops from the common border. Its dispute with Pakistan, over Kashmir, shows no sign of lessening and intensified in 1995, following the destruction of a mosque in Srinagar.

ECONOMY

A third of India's gross domestic product originates from the agricultural, mining, forestry and fishing sectors, and almost a third from industry. The main exports are gems, jewellery, clothing and engineering products.

Rao has produced something of a transformation in the economy. When he came to power he freed prices, reduced tariffs and made the currency partly convertible. As a result, inflation has fallen, the budget deficit has been reduced, foreign investment has increased dramatically and exports have expanded.

India now has to overcome its excessive bureaucracy and overmanning, but progress is being made. The rate of increase of the Indian population is now decreasing, which will lessen that burden on the economy.

THE REPUBLIC OF INDIA	
STATUS	Federal Republic
AREA	3,166,830 sq km (1,222,395 sq miles)
POPULATION	883,910,000
CAPITAL	New Delhi
LANGUAGE	Hindi, English, regional languages
RELIGION	83% Hindu, 11% Muslim
CURRENCY	Indian rupee (INR)
ORGANIZATIONS	Col. Plan, Comm., UN

PHYSICAL GEOGRAPHY
Indonesia, an archipelago of 13,677 islands, exhibits great physical diversity with tropical swamps, rainforest and over 300 volcanoes, many still active. Two-thirds of Indonesia's population lives on Java.

CLIMATE
The climate varies but tropical monsoons generally produce a wet season from October to April, with dry weather from June to September.

POLITICS AND RECENT HISTORY
For 27 years, since the downfall of President Sukarno, considered to be the architect of modern Indonesia, the country has been effectively ruled by President Suharto and the armed forces. The army's dual functions of defence and internal political control are enshrined in the constitution. The Suharto regime permits only three political parties. The government-run Golkar party is dominant and it has enjoyed a monopoly of power. The Independent Democratic Party may in future provide a more serious opposition, if only because it is led by Sukarno's daughter Megawati Sukarnoputri. However, there is much speculation as to whether Suharto will seek a further five-year term of office at elections due in 1997.

The Suharto government oversees a largely diverse and multi-ethnic nation and has developed the philosophy of *Pancasila* which, among other principles, promotes the deliberate protection of ethnic and religious distinction against an overriding background of national unity. Yet, despite this avowed toleration, Indonesia has stood accused of human rights abuses in Irian Jaya and Aceh province in Sumatra. Indonesia's annexation of Portuguese East Timor, still not recognized by the United Nations, has been prominent together with the well-publicized massacre of 100 civilians in that territory, by Indonesian troops, in 1991.

ECONOMY
The Indonesian economy has advanced consistently. Until the early 1980s, it was built predominantly on oil wealth. Although this commodity is still important, its relative significance to the economy has diminished as manufacturing industry has flourished. As oil reserves become less economic to exploit, Indonesia may in the next century become an oil importer. On the other hand, liquefied natural gas will retain its key role. Indonesia is the world's largest exporter, mainly to Japan.

Indonesia also enjoys considerable mineral wealth. It exports coal, and extremely valuable desposits of copper and gold are to be exploited with the help of foreign investment. The country is also a leading producer of forest products, palm oil, rubber and spices. It has, however, taken steps to conserve the forests of Kalimantan by tighter control of logging concessions.

Over the past 20 years, the Indonesian economy has grown steadily at an average rate of 6.8 per cent per annum. In 1970, 60 per cent of the population lived in poverty – the figure has now fallen to 15 per cent.

THE REPUBLIC OF INDONESIA	
STATUS	Republic
AREA	1,919,445 sq km (740,905 sq miles)
POPULATION	189,136,000
CAPITAL	Jakarta
LANGUAGE	Bahasa Indonesian, Dutch
RELIGION	88% Muslim, 9% Christian, Hindu and Buddhist minorities
CURRENCY	rupiah (IDR)
ORGANIZATIONS	ASEAN, Col. Plan, OPEC, UN

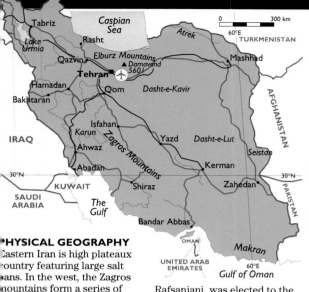

PHYSICAL GEOGRAPHY

Eastern Iran is high plateaux country featuring large salt pans. In the west, the Zagros mountains form a series of ridges, while to the north the Elburz mountains flank the southern shores of the Caspian Sea.

CLIMATE

Much of Iran, away from coasts, experiences extremes of climate with hot summers and bitterly cold winters, temperatures ranging from 20–55°C (-4–131°F). Most of the light rainfall is in winter and varies from 2,000 mm (99 inches) to almost zero.

POLITICS AND RECENT HISTORY

The overthrow of the Shah in January 1979 and the return a month later of Ayatollah Khomeini marked the beginning of the Iranian Islamic revolution, with power passing to the nation's spiritual leaders and isolationist policies prevailing.

Following an inconclusive but economically damaging war with Iraq, which ended with a cease-fire in 1987, and the death of Ayatollah Khomeini in 1989, the reformist former speaker of the Iranian parliament (*Majlis*), Hashemi

Rafsanjani, was elected to the presidency, an office he retained in elections in 1993. In recent years, the overall supremacy of Islamic fundamentalism has been curtailed, but the revolution continues. The clergy, through the nation's spiritual leader, Ayatollah Ali Khamenei, opposes change and presents the president with a barrier to political progress. The fatwa against the author Salman Rushdie is symbolic of this conflict.

There are signs that the Iranian government has been able to limit the most extreme authority of religious factions, as it seeks to achieve more normal relationships with other nations, but religious funda-

mentalism still wields substantial power.

ECONOMY

The Islamic revolution did not bring economic prosperity to Iran. Its effect, combined with the war with Iraq, was to reduce the nation from a state of relative prosperity to one of poverty. The damage was aggravated by the loss of some four million of the better-educated and more technically qualified population, who fled when the Shah was overthrown.

President Rafsanjani is clearly intent on free market reforms but is somewhat thwarted by the opposition of the clerics. Progress has been slow and is much hampered by the decline in oil prices. Demographic circumstances also take their toll – the population is increasing at a dramatic rate.

On the credit side, Iran is building a stronger economic relationship with Russia, who are to construct a nuclear power station for the Iranians. Iran is also taking the lead in arguing that those nations bordering the Caspian Sea should reach agreement on how to share its resources.

THE ISLAMIC REPUBLIC OF IRAN	
STATUS	Islamic Republic
AREA	1,648,000 sq km (636,130 sq miles)
POPULATION	60,270,000
CAPITAL	Tehran
LANGUAGE	Farsi, Kurdish, Arabic, Baluchi, Turkic
RELIGION	majority Shi'a Muslim. Sunni Muslim, Armenian, Christian minorities
CURRENCY	Iranian rial (IRR)
ORGANIZATIONS	Col. Plan, OPEC, UN

PHYSICAL GEOGRAPHY

The heart of Iraq is the lowland valley of the Tigris and Euphrates rivers. Northern Iraq is hilly, extending into the Zagros mountains, while western Iraq is desert.

CLIMATE

Summers are extremely hot with maximum temperatures between July and August usually exceeding 40°C (104°F). Winters are mild, with unreliable rainfall of less than 500 mm (20 inches) a year.

POLITICS AND RECENT HISTORY

Saddam Hussein came to power in the late 1970s and assumed the presidency when Ahmed al Bakr resigned in 1979. His authority was and is absolute; any expression of opposition is ruthlessly crushed and he relies on only a small clique of trusted associates. Through much of the 1980s, Iraq and Iran fought an inconclusive war, ended by a United Nations (UN) sponsored cease-fire in 1987. In August 1990 Iraq invaded Kuwait, precipitating the Gulf War and Iraq's absolute defeat at the hands of the coalition forces in March 1991. The UN cease-fire resolution imposed thereafter requires Iraq to destroy chemical and biological weapons, and to forgo the acquisition and development of nuclear weapons and ballistic missiles. It is enforced by economic sanctions and through a specially appointed commission.

After the Gulf War, Iraq continued its oppression of the Kurdish peoples in the north, and later mounted a campaign against the Shi'a people of southern Iraq, who inhabit the marshy regions of the lower Tigris and Euphrates. These events prompted western governments to declare a safe haven for Kurds in the north and the US, British and French governments to enforce a no-fly zone between 32°N and 36°N.

Despite these measures, Iraqi government oppression,

especially of the marsh Arabs, continues and Saddam Hussein's grip on power seems as firm as ever. Iraq has, however, asserted to the UN that it will recognize the sovereignty of Kuwait.

ECONOMY

UN sanctions have ruined the Iraqi economy. The prices of foodstuffs bear no relation to salaries, and the threat of serious famine looms. The black market economy has taken over but prices are out of reach of most of the population. UN resolutions, which would allow Iraq to sell some oil, provided that the proceeds were used for humanitarian needs and to pay reparations, have been rejected by Iraq as an infringement of sovereignty. The sanctions remain, mainly at the behest of the USA, on the grounds that Iraq has not met the UN conditions. However, international enthusiasm for the sanctions is waning, since they have not succeeded in the unstated objective of removing Saddam Hussein from power. Russia has reached an agreement to develop oilfields when sanctions are lifted and trade delegations from Britain and France have visited Baghdad.

THE REPUBLIC OF IRAQ	
STATUS	Republic
AREA	438,445 sq km (169,240 sq miles)
POPULATION	19,454,000
CAPITAL	Baghdād
LANGUAGE	Arabic, Kurdish, Turkoman
RELIGION	50% Shi'a, 45% Sunni Muslim
CURRENCY	Iraqi dinar (IQD)
ORGANIZATIONS	Arab League, OPEC, UN

PHYSICAL GEOGRAPHY

The Republic of Ireland is a lowland country of wide valleys, lakes and marshes. There are some hills of significance, especially in coastal regions, such as the Wicklow mountains south of Dublin, the Connemara mountains in the west and McGillicuddy's Reeks in the southwest.

CLIMATE

The Irish climate is maritime and influenced by the Gulf Stream. Rainfall is plentiful throughout the year and temperatures are mild.

POLITICS AND RECENT HISTORY

Albert Reynolds became prime minister (*Taoiseach*) of the Irish Republic when Charles Haughey resigned from office in January 1992, following allegations of financial irregularities. In elections a year later, Reynolds retained his premiership in parliament (the *Dáil*) in a coalition government in which the majority *Fianna Fáil* joined with the Labour Party. However, Reynolds was himself forced to resign in late 1994, when his attorney general

mishandled the extradition of a paedophile priest. A new coalition government was formed, led by John Bruton of *Fine Gael* with the support of Dick Spring's Labour party and the more radical Democratic Left party. Well before his resignation Reynolds, in co-operation with the British government, had sought to end the years of strife in Northern Ireland through the publication of a joint declaration to work toward peace. There was no immediate breakthrough, but eventually the IRA declared a cease-fire, signalling the beginning of political negotiations aimed at a lasting settlement. Bruton has shown no less urgency than his predecessor in pushing ahead with this process. In 1995, the British and Irish governments published a joint framework document on the future of Northern Ireland.

ECONOMY

The Irish economy is built predominantly on agriculture. The mild, moist climate is particularly suited to the farming of cattle, both for dairy produce and for meat, and of sheep. Arable crops include barley, oats and potatoes.

Membership of the European Union (EU) has been advantageous to the Irish nation, particularly its farmers who have benefited from heavy EU subsidies. The EU has also encouraged the development of manufacturing industries, and in this sector food processing, electronic and textile manufacturing has grown.

Discoveries of oil and gas off Ireland's southwest coast will further enhance prosperity, as will the exploitation of valuable lead and zinc reserves at Galmoy in the centre of the republic.

Ireland's main trading partner is the UK, but an increasing proportion of trade is with other EU nations. Ireland's economy, already growing fast, will be further boosted if lasting peace in Northern Ireland can be achieved, mainly by the benefit to tourism.

THE REPUBLIC OF IRELAND	
STATUS	Republic
AREA	68,895 sq km (26,595 sq miles)
POPULATION	3,571,000
CAPITAL	Dublin (Baile Átha Cliath)
LANGUAGE	Irish, English
RELIGION	95% Roman Catholic, 5% Protestant
CURRENCY	punt or Irish pound (IEP)
ORGANIZATIONS	Council of Europe, EEA, EU, OECD, UN

PHYSICAL GEOGRAPHY
Beyond the coastal plain of Sharon are a series of hills and valleys behind which is the Great Rift valley, formed from the valley of the river Jordan and the Dead Sea. Southern Israel, part of the Negev, is an arid plateau.

CLIMATE
The climate is generally Mediterranean with warm summers and mild winters, during which most rain falls. Southern Israel is hot and dry.

POLITICS AND RECENT HISTORY
Since the creation of the state in 1948, Israeli politics have been dominated by the struggle to maintain the nation's existence, conflict with Arab neighbours, and the issue of nationhood for displaced Palestinian peoples.

During the 1990s steps were initiated to achieve a peace agreement with the Palestinian Liberation Organization (PLO), which received a boost when the right-of-centre Likud Party was defeated by Labour at elections in July 1992. Yitzhak Rabin's new government was able to reach an accord with Yasser Arafat, the PLO leader, in 1993. Its provisions included the recognition of Israel by the PLO and a degree of autonomy for Palestinians in the occupied territories of Gaza and the West Bank of the Jordan. Initially the autonomy of the West Bank was limited to Jericho. The murder of Rabin in November 1995 proved disruptive to the peace process, and although peace talks continue, progress has been negligible. The Israeli government, under pressure from its right wing and Orthodox religious opponents, finds it convenient to demand that Yasser Arafat control terrorism. He in

turn argues that he seeking some sign of Israeli movement on the West Bank. Despite the lack of progress, Israel's policy is clearly to work towards peace with its neighbours and there are now signs of possible negotiations with Syria.

ECONOMY
The Israeli economy has always been influenced by the large defence budget, and this exerts pressure to achieve peace. The nation has a strong industrial base encompassing the range from heavy manufacturing, for example of steel, to high-tech light industry. Agriculture, based on the Kibbutz system, is also important, and citrus fruits provide valuable export revenue.

Energy sources are of strategic importance to Israel, although some oil is extracted in the Negev. In 1993 it was announced that natural gas would be used for electricity generation and, more recently, Israel concluded a deal with Egypt for the supply of gas. Trading partnerships with of vital importance – the country has trading relationships with Europe, China and a long-standing, close association with the USA.

THE STATE OF ISRAEL	
STATUS	State
AREA	20,770 sq km (8,015 sq miles)
POPULATION	5,423,000
CAPITAL	Jerusalem
LANGUAGE	Hebrew, Arabic, Yiddish
RELIGION	85% Judaism, 13% Muslim
CURRENCY	shekel (ILS)
ORGANIZATIONS	UN

PHYSICAL GEOGRAPHY

The dominant physical feature of northern Italy is the Alps, which merge with the Apennines to form the spine of Italy. Southern Italy is intermittently volcanically active.

CLIMATE

The Italian climate is typically Mediterranean, enjoying warm summers and mild winters with some rain.

POLITICS AND RECENT HISTORY

For many years, Italian politics were generally dominated by a series of coalitions from the centre of the political spectrum. Gradually, the centre became discredited for its support of what was seen as a corrupt system. This eclipse of power came to fruition at the March 1994 general election, when the media tycoon Silvio Berlusconi, heading his newly formed *Forze Italia* party, formed a right wing coalition government with the support of the federalist Northern League and the neo-fascist National Alliance. By the end of 1994, Berlusconi, accused of corrupt practices, lost the support of the Northern League who brought a vote of no confidence, and he was forced to resign. After a difficult few weeks, President Scalfaro was able to persuade a banker, Lamberto Dini, to form the 54th post-war Italian administration. This is regarded as a caretaker government, with ministerial appointments from academia and the business world.

Despite this turbulence, the Italian campaign against corruption generally, and the Mafia in particular, continues.

ECONOMY

Italy has a fundamentally strong economy based on agriculture, mainly cereals, wine, fruit and vegetables, and industry in which iron and steel, chemicals, textiles and car manufacturing are important. Among service industries tourism is a major source of revenue.

The country is now emerging from recession, mainly through a growth in exports, but the country has been plagued by large budget deficits. As a result, automatic indexation of pay rises has ended and it seems likely that the country's generous pension arrangements will have to be reviewed.

Italy fought hard to stay within the European exchange rate mechanism but, following a 7 per cent devaluation of the lira, was unable to sustain membership. Since then, the currency has suffered further significant losses in its value.

THE REPUBLIC OF ITALY	
STATUS	Republic
AREA	301,245 sq km (116,280 sq miles)
POPULATION	57,110,000
CAPITAL	Rome (Roma)
LANGUAGE	Italian, German, French
RELIGION	90% Roman Catholic
CURRENCY	Italian lira (ITL)
ORGANIZATIONS	Council of Europe, EEA, EU, G7, NATO, OECD, UN, WEU

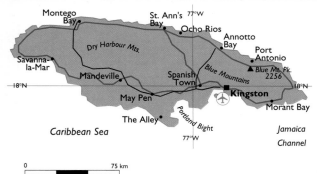

PHYSICAL GEOGRAPHY

Jamaica is the third largest island in the Caribbean. It has a central mountain range, rising to Blue Mountain Peak at 2,256 m (7,400 ft) in the east. Low coastal plains are interrupted by hills and plateaux.

CLIMATE

The climate is tropical, but temperate in the mountain regions. The rainy season lasts from May to October. Jamaica is very prone to hurricanes.

POLITICS AND RECENT HISTORY

In 1962 Jamaica achieved full independence within the Commonwealth. It has two main political parties, the Jamaica Labour Party (JLP), led by Edward Seaga, and the People's National Party (PNP). The latter, led by Percival J. Patterson, won a general election in March 1993. The JLP boycotted parliament until July 1993 to show its dislike of electoral reforms and alleged electoral malpractice by the PNP.

ECONOMY

Jamaica is the world's third largest producer of bauxite,

after Australia and Guinea. At current rates of extraction, the proven reserves could last 200 years. Low world bauxite prices and falling international demand have hampered economic growth, since this sector is responsible for 60 per cent of Jamaica's exports. The aim of the industry now, is to export less and refine more. The alumina refineries are being expanded and refurbished to increase output. Plans to link an aluminium smelting industry with Trinidad and Tobago, however, have been given lower priority because of the state of the market.

Agriculture plays a major part in the economy, producing bananas, cocoa, coffee and especially sugar cane, which is exported to the USA and the European Union under preferential quotas. After Hurricane Gilbert struck in September 1988, causing immense destruction and damage, the

cash crops took nearly three years to recover. Floods in 1993 severely affected the sugar and banana crops.

Within the last decade, tourism, mainly from the USA, has become the major source of foreign earnings, especially welcome while markets for Jamaica's bauxite, bananas and sugar are weak. A recent trend has been a decline in the hotel sector resulting from a surge in cruise ship activities, while further competition to the tourist industry is emerging as new resorts are developing fast, such as those in the Dominican Republic, Cuba and Mexico.

Remittances from Jamaicans living abroad, particularly in the USA, contribute significantly to foreign exchange earnings and help to alleviate poverty. High inflation in the 1990s has been a critical issue in Jamaica, resulting in a lowering of living standards and increasing an already wide gap between the rich and poor.

JAMAICA	
STATUS	Commonwealth State
AREA	11,425 sq km (4,410 sq miles)
POPULATION	2,411,000
CAPITAL	Kingston
LANGUAGE	English, local patois
RELIGION	Anglican Christian majority, Rastafarian minority
CURRENCY	Jamaican dollar (JMD)
ORGANIZATIONS	Caricom, Comm., OAS, UN

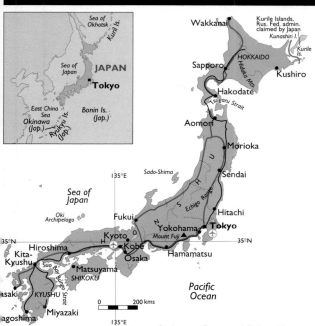

PHYSICAL GEOGRAPHY

Japan consists of four major islands, Hokkaido, Honshu, Shikoku and Kyushu, which stretch over 1,600 km (995 miles). Eighty per cent of the population live on the central island of Honshu. There are more than 3,000 smaller volcanic mountainous islands and over 60 active volcanoes, subject to frequent major earthquakes, monsoons, typhoons and tidal waves. Nearly three-quarters of the land is mountainous and is generally heavily forested, with small, fertile areas. The highest point is Mt. Fuji, which reaches a height of 3,776 m (12,388 ft).

CLIMATE

The climate is temperate oceanic, with warm summers and mild winters, except in western Hokkaido and northwest Honshu, where the winters are very cold with heavy snowfall. Rain falls mainly between June and July with typhoons sometimes occurring during September.

POLITICS AND RECENT HISTORY

Political events in Japan in recent years have been dominated by scandals affecting the nation's leading politicians. In 1989 the resignation of the prime minister, Noboru Takeshita, over a matter of financial impropriety, and the resulting reverberations threatened the Liberal Democratic Party's (LDP) hold on power for the first time since 1955.

JAPAN	
STATUS	Constitutional Monarchy
AREA	369,700 sq km (142,705 sq miles)
POPULATION	124,536,000
CAPITAL	Tokyo
LANGUAGE	Japanese
RELIGION	Shintoism, Buddhism, Christian minority
CURRENCY	yen (JPY)
ORGANIZATIONS	Col. Plan, G7, OECD, UN

Successive prime ministers, Sosuke Uno, Toshiki Kaifu and Kiichi Miyazawa came and went, the last when his plans to combat corruption were scuppered by party colleagues. At elections in 1993, the LDP conceded power after 88 years in office and a coalition was formed under Morihiro Hosokawa. His government lost the support of the Socialists and an even more fleeting government, under Tsutomu Hata, took office in early 1995. By June 1995, his coalition had failed and the Socialist Tomiichi Murayama formed a coalition, albeit with substantial Liberal Democrat support. Shortly afterwards, three former prime ministers, Hosokawa, Hata and Kaifu, announced the formation of the New Frontier Party to oppose the incumbents. Following the slow response to the Kobe earthquake and the poisoned gas attack on the Tokyo metro, it became clear that Murayama's coalition was losing the support of the Japanese population. A degree of uncertainty in Japanese internal politics for the foreseeable future seems likely.
Internationally, Japan has

now gone further than ever before in apologizing for war atrocities, but has not agreed to compensation payments. The ban on military activity is weakening and small contingents have been sent in support of United Nations operations in Cambodia, Mozambique and Rwanda. There is a wish to finally sign a peace treaty with Russia, but Japan will not do this until a satisfactory resolution concerning the future of the southern Kurile Islands, occupied at the end of World War II, is reached.

ECONOMY

The resurgence of the Japanese economy out of the disaster of World War II is truly astounding. Japan has the world's largest industrial economy, despite the fact that there are few natural resources and most raw materials have to be imported, including around 90 per cent of energy requirements. To reduce this dependence, the country is developing nuclear power resources and increasing the production of its limited resources of coal, oil and natural gas.

By the late 1980s, the country had become the world's largest manufacturer of cars, many consumer durables such as washing machines, and instruments such as watches and calculators. The nation was at the forefront of the electronics industry. Japanese success in such industries spawned a degree of resentment overseas that encouraged governments in Europe and the United States to impose tariff barriers. The Japanese response was to invest heavily in productive facilities overseas, especially in the European Union. One curious by-product of this policy is that in 1993 Japan became a net importer of cars.

Most food has to be imported but the Japanese catch and eat a lot of fish. The fishing fleet is the largest in the world.

The recessionary years of the 1990s have effected the Japanese economy and although it is recovering, growth has been slow in contrast to the earlier spectacular development. To some extent, Japan is the victim of its own success. Its large trade surplus has led to the continuing rise in the value of the yen. Japanese businesses are beginning to find that they are no longer as price competitive as they once were. Even so, Japan remains economically strong with a powerful work ethic and low unemployment, although the government may be forced to lift the restrictions on foreign involvement in domestic commerce and industry.

Japan continues to invest in its infrastructure, the latest example being the Tokyo Bay highway, which, when completed, will be the world's widest under-sea tunnel.

HOKKAIDO

STATUS	Island of Japan
AREA	78,460 sq km (30,285 sq miles)
POPULATION	5,644,000
CAPITAL	Sapporo

HONSHU

STATUS	Main Island of Japan
AREA	230,455 sq km (88,955 sq miles)
POPULATION	99,254,000
CAPITAL	Tokyo

KYUSHU

STATUS	Island of Japan
AREA	42,010 sq km (16,215 sq miles)
POPULATION	13,296,000
CAPITAL	Fukuoka

SHIKOKU

STATUS	Island Prefecture of Japan
AREA	18,755 sq km (7,240 sq miles)
POPULATION	4,195,000
CAPITAL	Matsuyama

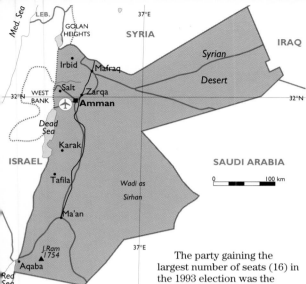

PHYSICAL GEOGRAPHY

Most of the land area of Jordan to the east and southeast is desolate plateaux with occasional salt pans. The lowest parts are along the eastern shores of the Dead Sea and the East Bank of the Jordan, behind which is a range of tree-covered hills.

CLIMATE

Most of Jordan is hot and dry with temperatures rising to 49°C (120°F) in the valleys. The west of the country is cooler and wetter with rainfall of up to 290 mm (12 inches).

POLITICS AND RECENT HISTORY

Jordan is a constitutional monarchy in which King Hussein, who succeeded to the throne in 1952, wields considerable power. Nevertheless, in November 1993 the first free multi-party elections to the 80-member parliament were held. The previous elections in 1989 were free but political parties were illegal and all candidates stood as independents.

The party gaining the largest number of seats (16) in the 1993 election was the Islamic Action Front, representative of fundamentalist views. However, this revealed a decline from the previous parliament and the Front will not be able to interfere too much with King Hussein's policies. Thus, the King's decision to include Islamic fundamentalist factions in the political process has been vindicated.

Following the Israeli–Palestine Liberation Organization (PLO) agreement reached in September 1993, Jordan, although displeased at exclusion from the negotiations, has been seeking to reach a peace accord with Israel. By the end of 1993 it was reported that agreement was near but that Jordan was

THE HASHEMITE KINGDOM OF JORDAN	
STATUS	**Kingdom**
AREA	**90,650 sq km (35,000 sq miles)**
POPULATION	**4,936,000**
CAPITAL	**Amman**
LANGUAGE	**Arabic**
RELIGION	**90% Sunni Muslim, Christian and Shi'ite Muslim minority**
CURRENCY	**Jordanian dinar (JOD)**
ORGANIZATIONS	**Arab League, UN**

still seeking access to the Israeli and Palestinian economies as part of the deal. In January 1994, a Jordan-PLO accord allowed Jordanian banks on the West Bank, closed since 1967, to re-open.

ECONOMY

Jordan is dependent upon substantial injections of foreign aid for its economic survival. It has incurred serious debts both in its balance of payments and through its internal budget deficit. The continued flow of aid has been dependent on the stability of the Jordanian government but was somewhat threatened by the equivocal stance that the nation took during the Gulf War.

Since then attempts have been made to restore good relations, especially with the Gulf States. There is also an economic incentive in achieving a peace accord with Israel which is likely to result in some relief from Jordan's debt.

The Jordanian economy, which is largely agricultural, was severely damaged by the loss of its West Bank territories in the Six-Day War. On its remaining cultivable land the principal crops are a variety of fruits and vegetables, such as tomatoes, cucumbers, melons and citrus fruits.

PHYSICAL GEOGRAPHY

The republic covers a vast area of Central Asian steppe, verging towards semi-desert in the south. The terrain is lowest around the Caspian Sea and highest in the mountains to the east.

CLIMATE

The climate is essentially dry, with a marked seasonal variation in temperature.

POLITICS AND RECENT HISTORY

Kazakhstan achieved international recognition as an independent nation at the end of 1991. It is a presidential republic and, in late 1991, Nursultan Nazarbayev was elected unopposed. Some political reform is being pursued and a new constitution has been adopted. In March 1995, President Nazarbayev dismissed the elected parliament, on the grounds that the election result was not valid, and announced rule by decree. The following month he obtained an overwhelming majority in a plebiscite, allowing him to retain office until the year 2000.

There are some signs of discord between the ethnic Kazakh population and

Russians within the republic. Kazakhstan needs close ties with Russia – the two countries have agreed to unite their armies and, in a move to mollify anxious Russian inhabitants, the President has proposed moving the capital to the northern city of Akmola.

ECONOMY

Kazakhstan has rapidly embraced the concept of economic reform and the creation of a market economy. It has established free economic zones, with tax advantages, to attract foreign capital. State farms were privatized by decree, although many farmers have found private ownership unrewarding. The government has left the rouble zone and the tenge is the legal currency.

Kazakhstan has immense

THE REPUBLIC OF KAZAKHSTAN	
STATUS	Republic
AREA	2,717,300 sq km (1,048,880 sq miles)
POPULATION	16,925,000
CAPITAL	Alma-Ata
LANGUAGE	Kazakh, Russian
RELIGION	Muslim majority, Orthodox minority
CURRENCY	tenge
ORGANIZATIONS	CIS, UN

mineral resources, with massive deposits of iron ore and important reserves of lead, zinc, titanium, chromium, vanadium, gold and silver. Thallium vital for the electronics industry, is found at a purity unrivalled elsewhere. Close to the Caspian Sea, large oil and natural gas reserves are being exploited with the help of foreign investment.

Most agriculture is pastoral Across the steppes, wheat is grown, and further south vegetables, cotton and tobacco are grown on irrigated land. This has focused attention on the one major commodity in short supply – water. Drinking water is piped long distances to many settlements. The pipes are old and replacement will require substantial investment. Irrigation in the south has used the waters from the Aral Sea, which has drastically diminished in size, leaving many square miles of salt pan and creating, in the eyes of some, an ecological disaster.

Despite economic progress output has not yet reached pre-independence levels, inflation is high and there are signs of industrial unrest, particularly in the coal mines, where wages have not been paid. Despite these difficulties, long-term economic prospects are good.

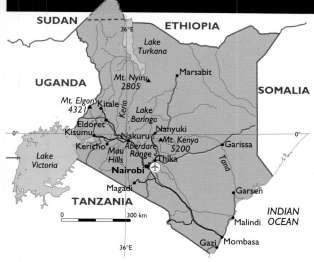

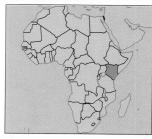

PHYSICAL GEOGRAPHY

The land rises from the coastal plains to plateaux of mainly volcanic material, at altitudes generally in excess of 1,500 m (5,750 ft). Running northwest of Nairobi to Lake Turkana is the dramatic Great Rift Valley. To the east of the valley rise the Aberdare mountains and, further east, Mt. Kenya at 5,200 m (17,060 ft) high.

CLIMATE

Although Kenya straddles the Equator, only the coastal fringes exhibit tropical characteristics because of the high average altitude. Temperatures range from 10–27°C (50–81°F). Much of the land north of the equator is arid semi-desert, while to the south savannah conditions prevail. Rainfall varies from 760–2,500 mm (30–98 inches) depending on altitude.

POLITICS AND RECENT HISTORY

Kenya achieved independence from Britain in 1963 and in 1964 became a republic. The first president was Jomo Kenyatta, who led the unrest that had preceded independence.

Kenyan politics are dominated by tribal allegiance in which the Kikuyu, the majority tribe, has until recent years been ascendant. Although recent political history is apparently one of stability, the government is not truly democratic. Multi-party elections, the first for nearly 25 years, were held in December 1992, following pressure from Britain, France and other western creditors. The Kenyan African National Union (KANU) and President Daniel arap Moi retained power, but the election was not regarded as entirely free and fair by observers. Since then, ethnic violence has continued. Masai people have driven Kikuyu from their homes, apparently with the connivance of the government. There is also evidence of genocide of Turkana people in northern Kenya. The announcement of a new political party, by Paul Muite and Dr Richard Leakey, the former Director of the Kenyan Wildlife Service, has been attacked by KANU.

ECONOMY

Kenya's economy is founded on agriculture and tourism, with coffee and tea forming the most important export commodities.

The country's prosperity declined in the early 1990s – economic growth fell and foreign debt escalated. However, President Moi's government followed the advice of the International Monetary Fund and the World Bank (who at one time suspended aid) and have convinced these agencies of the country's financial rectitude. The future of the economy is, nevertheless, uncertain and it has recently suffered from a severe drop in tea production.

THE REPUBLIC OF KENYA	
STATUS	Republic
AREA	582,645 sq km (224,900 sq miles)
POPULATION	28,113,000
CAPITAL	Nairobi
LANGUAGE	Kiswahili, English, Kikuyu, Luo
RELIGION	majority traditional beliefs, 25% Christian, 6% Muslim
CURRENCY	Kenya shilling (KES)
ORGANIZATIONS	Comm., OAU, UN

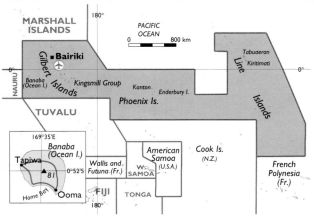

PHYSICAL GEOGRAPHY

Kiribati consists of the volcanic island of Banaba and three groups of coral atolls spread over five million sq km (1,930,000 sq miles) in the central and west Pacific: 16 Gilbert islands, eight Phoenix islands and eight of the Line islands. The highest point, on Banaba, reaches 81 m (265 ft).

CLIMATE

The central islands have a maritime equatorial climate, whereas those to the north and south are tropical, with constant temperatures of 27–32°C (80–90°F). The highest rainfall occurs in the northern islands.

POLITICS AND RECENT HISTORY

Kiribati was formerly the Gilbert Islands, part of the British colony of the Gilbert and Ellice Islands. In 1975 the Ellice Islands seceded as a separate territory (becoming Tuvalu) and in 1979 the remaining islands became the independent Republic of Kiribati. The president is head of state and shares executive power with the government. The main islands are heavily populated, and a resettlement scheme involves the dispersal of 4,700 people from the overcrowded islands to other atolls.

Kiribati is a member of the South Pacific Forum, an informal gathering of government leaders from states in the South Pacific who meet annually, or when the need arises, to discuss issues of common concern. The Forum has adopted a treaty for a nuclear-free zone in the region, prohibiting the testing or possession of nuclear weapons. Other issues have included fishing rights and concern over the use of drift nets, which cause the depletion of fish stocks and other marine life. The fishing nations concerned were confronted and restrictions on fishing were imposed. Discussions in recent years have focused on global warming and the threat to the region of a predicted world rise in sea levels. This could overwhelm the low-lying islands within Kiribati, Tuvalu and Tonga, and it has been agreed that monitoring stations will be set up throughout the region to gauge any climatic change.

ECONOMY

Before 1980, the economy was based on phosphates from Banaba (formerly Ocean Island), the export of which provided 85 per cent of the group's revenue – a fact which encouraged the Banabans to make an unsuccessful bid for independence on their own. When mining on the island ceased the economy collapsed. Kiribati turned to coconuts, and copra is now the major export. However, this is subject to variations in international copra prices. Fish, in particular tuna, account for one third of exports, and further income is derived from licences issued to foreign fishing fleets, chiefly from the USA, Japan, Taiwan and South Korea.

THE REPUBLIC OF KIRIBATI	
STATUS	Republic
AREA	717 sq km (277 sq miles)
POPULATION	77,000
CAPITAL	Bairiki on Tarawa Atoll
LANGUAGE	I-Kiribati, English
RELIGION	Christian majority
CURRENCY	Australian dollar (AUD)
ORGANIZATIONS	Comm., UN

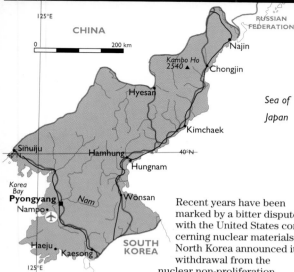

PHYSICAL GEOGRAPHY
Most of North Korea is rugged mountainous terrain. The greatest extent of lowland is to the south and west of Pyongyang.

CLIMATE
Temperatures range widely from a summer maximum of above 30°C (86°F) to a winter minimum of -10°C (14°F). Most rain falls in the summer months.

POLITICS AND RECENT HISTORY
North Korea, separated from its southern neighbour at the end of World War II, is one of the few remaining strongholds of Communist totalitarianism. President Kim Il Sung (the Great Leader) came to power in 1948 and ruled until his death in 1994. He was succeeded by his son Kim Jong Il, known as Dear Leader. Like his father, Kim Jong Il wields virtually absolute power, but the new generation of leadership may allow more openness. Certainly, the Korean population, hungry and repressed, has become increasingly impatient with the leadership.

Recent years have been marked by a bitter dispute with the United States concerning nuclear materials. North Korea announced its withdrawal from the nuclear non-proliferation treaty, and was suspected of stock-piling plutonium and believed capable of building a nuclear bomb. Ex-president of the USA, Jimmy Carter, mediated and negotiations produced an agreement in which two light-water reactors will be provided to North Korea at a cost of $4 billion. In turn, North Korea will freeze its own nuclear capability. So far the deal has not reached fruition, as North Korea objects to South Korea's role in providing the reactors.

ECONOMY
The North Korean economy has suffered grievously since losing support from Russia. Even China, a previously steadfast ally, no longer offers favoured nation treatment. As a result the nation's economy has contracted. Its large, inefficient industrial complexes operate well below capacity, and manufacturing output has fallen. Mining and agriculture have also declined. A major support of the economy is the cash, generated from pinball machines, and brought in by North Korean exiles living in Japan.

The government is finding it increasingly difficult to import sufficient oil, and there is now an acute shortage. The shortage of rice has compelled the North Koreans to seek help from Japan, who have agreed to provide contingency aid. A trade delegation from Japan has visited North Korea, possibly a sign that some measure of economic reform is being contemplated.

THE PEOPLE'S DEMOCRATIC REPUBLIC OF KOREA	
STATUS	Republic
AREA	122,310 sq km (47,210 sq miles)
POPULATION	23,048,000
CAPITAL	Pyongyang
LANGUAGE	Korean
RELIGION	Chundo Kyo, Buddhism, Confucianism, Daoism
CURRENCY	North Korean won (KPW)
ORGANIZATIONS	UN

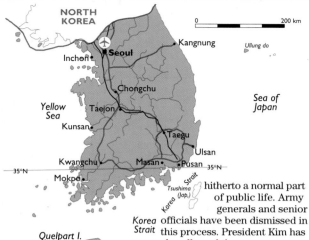

hitherto a normal part of public life. Army generals and senior officials have been dismissed in this process. President Kim has also allowed the press more freedom, but has, nevertheless, closed several newspapers.

To some extent, President Kim's attempts to introduce a greater measure of democracy have been frustrated by tensions with North Korea, caused principally by that country's intransigence over nuclear non-proliferation obligations. Although these tensions have reduced, they have reinforced the autocratic instincts of conservative elements in South Korea and made the process of democratization and political reform harder to achieve.

PHYSICAL GEOGRAPHY
South Korea is a country of rugged, mountainous terrain. The flattest and most populous parts lie along the west coast and in the extreme south of the peninsula.

CLIMATE
Summers, when most rainfall occurs, are hot with daytime temperatures approaching 30°C (86°F), while winters are often bitterly cold.

POLITICS AND RECENT HISTORY
South Korea, a product of territory occupied by US troops at the end of World War II, has experienced military dictatorship for most of its history. A transitional period under the presidency of former general Roh Tae Woo came to an end in February 1993 when Kim Young Sam, of the Democratic Liberal Party, was elected to the presidency.

Kim is the first civilian head of state in 32 years. He has concentrated on the advancement of democratic principles and, in particular, has campaigned against bribery and corruption,

ECONOMY
South Korea has few natural resources apart from coal and tungsten. Traditional industries include textiles, chemicals, shipbuilding and vehicle manufacture. In recent years the country has become known for specializing in electronics and computers.

Since taking office, Kim's priority has been to deregulate the economy. He has introduced legislation aimed at banning financial malpractice, such as bank accounts in false names, and proposes the privatization of 23 government-controlled businesses. South Korea has applied to join the Organization for Economic Co-operation and Development.

The South Korean economy has shown remarkable success over the past 30 years and the country is, by any standard, massively more prosperous than its northern neighbour. There are some tentative moves towards economic co-operation with the North, at this stage confined to light industry.

South Korea, for many years a borrower from the World Bank and International Monetary Fund, is now one of the elite group of lenders. However, prosperity has a price and recent times have been marked by industrial troubles, particularly at the giant Hyundai Corporation.

THE REPUBLIC OF KOREA	
STATUS	Republic
AREA	98,445 sq km (38,000 sq miles)
POPULATION	44,619,000
CAPITAL	Seoul (Sŏul)
LANGUAGE	Korean
RELIGION	26% Mahayana Buddhist, 22% Christian. Confucianism, Daoism, Chundo Kyo
CURRENCY	won (KRW)
ORGANIZATIONS	Col. Plan, UN

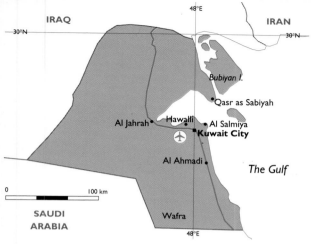

PHYSICAL GEOGRAPHY
Kuwait is essentially low-lying desert terrain with the lowest land in the north. Cultivation is only possible along the coast.

CLIMATE
In winter temperatures are mild and little rain falls. By contrast, summer months are very hot and dry with maximum daily temperatures reaching 40°C (104°F). Annual rainfall fluctuates between 10–370 mm (0.5–15 inches).

POLITICS AND RECENT HISTORY
Kuwait is a monarchy in which the Amir is selected from among the ruling dynasty, the Al Sabah family. The present ruler succeeded at the end of 1977. The National Assembly, which had been dismissed by the Amir in 1986, was reconstituted after the Gulf War, due to popular pressure, and has shown a measure of independence. Although real political power remains vested in the Al Sabah dynasty, the rulers will probably have to accept some devolution of authority to the Assembly. As yet, however, political parties are not permitted.

The political scene in Kuwait has naturally been dominated by the aftermath of the Gulf War. Iraq succeeded in heightening the tension once more when, in October 1994, it massed troops along the Kuwait/Iraq border. It withdrew under international pressure and subsequently, following mediation from Moscow, announced that it had abandoned its claim to Kuwaiti territory and would henceforth respect Kuwait's borders. Nevertheless, Kuwait's political priority is its defence and major agreements have been concluded with the USA, UK and France. Similarly, a pact with Russia provides arrangements for training and possible purchase of arms. More recently, Kuwait has agreed a closer relationship with Saudi Arabia, to counter possible future threats from Iraq or Iran.

ECONOMY
The Kuwaiti economy is dominated by its oil wealth. The oilfields have now largely recovered from the devastation of the Gulf War, and revenue from oil and petroleum products amounts to about 90 per cent of exports.

Even so, especially with low oil prices, Kuwait has experienced budget deficits. In part, this is attributable to the huge reconstruction costs after the war. However, it is unlikely that oil wealth will ever again give the Kuwaiti population the comfortable, insulated life style it once did. The country will, in all probability, have to allow foreign participation in the economy and may have to levy tax from its citizens.

THE STATE OF KUWAIT	
STATUS	State
AREA	24,280 sq km (9,370 sq miles)
POPULATION	1,433,000
CAPITAL	Kuwait (Al Kuwayt)
LANGUAGE	Arabic, English
RELIGION	95% Muslim, 5% Christian or Hindu
CURRENCY	Kuwaiti dinar (KWD)
ORGANIZATIONS	Arab League, OPEC, UN

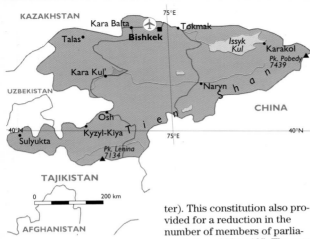

PHYSICAL GEOGRAPHY

Kyrgyzstan, occupying the western end of the Tien Shan range, is one of the world's most rugged and mountainous areas. Seventy per cent of its area is covered by glaciers or permanent snow. The Fergana valley in the west and the Chu valley in the north are the main areas of cultural and economic development.

CLIMATE

The mountainous areas have a hostile climate. In the lower valleys, summers are generally warm and winters cold.

POLITICS AND RECENT HISTORY

President Askar Akayev was first appointed in 1990. In August 1991 he withstood attempts by old-guard Communists to depose him and two months later the republic was declared independent from the former Soviet Union. In free elections, President Akayev was re-elected and given a democratic mandate. He has pursued political reform and was the architect of a new constitution which separated the head of state (the president) from the head of government (the prime minis-

ter). This constitution also provided for a reduction in the number of members of parliament, from 350 to 105. The constitution was approved by referendum, and elections to the new parliament were held in early 1995.

Kyrgyzstan has sought to establish relations with other nations, particularly Turkey with whom it shares linguistic traditions. The republic seeks to introduce democracy in line with the Turkish model. In its pursuit of a national identity, Kyrgyzstan has alienated some of the Russian population, many of whom have left the country, by its refusal to accept Russian as an official language.

ECONOMY

Kyrgyzstan is set firmly on the route of modernizing its economy through privatization and by attracting foreign investment. It has sought to establish business links with market-orientated nations and has, for

example, set up a joint venture with a South Korean company to produce consumer electronic goods. Its economy has been supported by the International Monetary Fund and the World Bank, from whom it has earned praise for its progress. The country has ambitions to become a provider of financial services within the region.

Kyrgyzstan has valuable mineral deposits, including gold, silver, mercury and antimony and rare earth metals such as yttrium and lanthanum. Oil has been found, reportedly in significant quantities, in the Ak Sui valley in the east but investment appraisal is necessary to determine whether it is exploitable.

Despite this, the Kyrgyzstan economy is essentially agricultural, based on livestock farming of sheep and goats, and also sericulture, bee keeping and tobacco growing.

Kyrgyzstan has withdrawn from the rouble zone and its currency is reasonably strong. The government has imposed spending cuts in order to bring the budget into balance, although increased taxation will probably be necessary.

THE KYRGYZ REPUBLIC	
STATUS	Republic
AREA	198,500 sq km (76,620 sq miles)
POPULATION	4,474,000
CAPITAL	Bishkek
LANGUAGE	Kyrgyzian, Russian
RELIGION	Muslim
CURRENCY	som
ORGANIZATIONS	CIS, UN

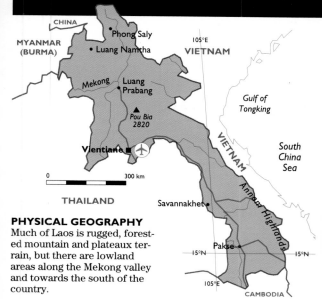

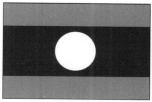

PHYSICAL GEOGRAPHY

Much of Laos is rugged, forested mountain and plateaux terrain, but there are lowland areas along the Mekong valley and towards the south of the country.

CLIMATE

In general the climate is hot with temperatures averaging 15°C (59°F) in winter and 32°C (90°F) before the rains in May–October. During the rainy season temperatures fall to around 26°C (79°F).

POLITICS AND RECENT HISTORY

Formerly a French protectorate, Laos became a constitutional monarchy in 1949 and a republic in 1975.

In common with its Indo-Chinese neighbours, Laos has suffered the turbulence of armed conflict within its borders. The Communist government of the 1970s, strongly influenced by Vietnam, remains in power but Vietnamese influence has waned. Laos is a one-party state ruled by the Politburo of the Laotian People's Party (LPRP) with the Pathet Lao as its military wing. Nouhak Phomsavan was elected president by the National Assembly in 1993.

Despite its Communist heritage, the Laotian government has in recent years embarked upon a reform programme and

has formed much closer ties with neighbouring Thailand. However, it remains to be seen whether the LPRP will introduce true democracy and allow opposition parties to operate legally.

ECONOMY

Laos remains one of the world's poorest nations, with an economy based primarily on subsistence farming. However, in seeking conversion to a market economy the government has set a course for at least a modest increase in prosperity. Greatest priority is being placed on infrastructure projects, on which 40 per cent of the slim budget is being expended. The road network is being

THE LAO PEOPLE'S DEMOCRATIC REPUBLIC	
STATUS	Republic
AREA	236,725 sq km (91,375 sq miles)
POPULATION	4,605,000
CAPITAL	Vientiane (Viangchan)
LANGUAGE	Lao, French, English, tribal languages
RELIGION	Buddhist majority, Christian and animist minorities
CURRENCY	kip (LAK)

improved and the recent opening of a bridge across the Mekong between Laos and Thailand, built with Australian aid, completes a highway between Singapore and Yunnan province in China. In late 1994 Laos cooperated with Cambodia, Vietnam and Thailand in plans for developing the Mekong River basin.

Agreements with Thailand are allowing the development of hydro-electric power, surpluses from which will be sold to Thailand, which presently provides 35 per cent of total overseas investment. Laos currently produces 120 MW of hydro-electric power, but is thought to have the potential for generating 20,000 MW. A programme to build 20 new dams is envisaged by the government and an Australian company has already embarked on a $6 billion, 600 MW dam project.

Laos has natural resources in the form of mineral ores, coal and timber, but commercial exploitation is overshadowed by the trade in opium and illegal logging activities. Nevertheless, with the help of overseas aid and foreign investment, the Laotian economy seems set to improve.

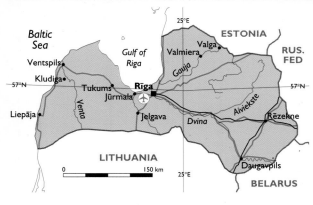

PHYSICAL GEOGRAPHY

The landscape is one of glaciated lowland, flat towards the Baltic coast but rather more hilly inland, with woods and lakes.

CLIMATE

The climate is one of distinct seasonal variation. Summers are warm but winters are cold, with average temperatures well below freezing. Precipitation occurs throughout the year, often as snow in winter.

POLITICS AND RECENT HISTORY

The campaign for Latvian independence began in earnest in 1990, when the Popular Front of Latvia won electoral victory. In an attempted Communist coup, Russian troops seized public buildings but were repulsed in January 1991. In August of that year independence was declared and internationally recognized.

By June 1993, when the first post-independence elections were held for a new 100-seat parliament, the authority of the Popular Front had dissipated and it was unable to secure enough of the vote to claim a single seat. The moderate nationalist party, known as the Latvian Way, gained most seats and has formed various coalitions under several prime ministers.

A dominant issue in Latvian politics has been citizenship. Initially, parliament enacted laws that granted citizenship only to ethnic Latvians, disenfranchising the substantial Russian minority. However, faced with international pressure and the threat from Russia's armed forces within Latvia, the government relented. Knowledge of the Latvian language, which most Russians have not learnt, remains a qualification for citizenship.

ECONOMY

Latvian economic policy is geared towards making a successful transition from the status of a republic within the former Soviet Union, to genuine independence. This is a difficult process which creates stresses. Working in concert with the International Monetary Fund, the government has introduced tight monetary policies to curtail government spending. Immediately following independence, inflation soared and the value of the Latvian rouble collapsed. Inflation is now under control and the new currency is stable.

The government is working hard to develop markets in the West, to replace those it has lost in the East. It has an agreement with Lithuania to build an oil terminal at the Latvian port of Liepaja, so that both countries will be less dependent on supplies from Russia.

Traditionally, the Latvian economy is oriented towards agriculture, and livestock farming, mainly of beef and dairy cattle and pigs, is important. Manufacturing industry is varied but specialist, concentrating on a few products such as radios, tape recorders, farm machinery and minibuses.

THE REPUBLIC OF LATVIA	
STATUS	Republic
AREA	63,700 sq km (24,590 sq miles)
POPULATION	2,536,000
CAPITAL	Riga
LANGUAGE	Latvian, Lithuanian, Russian
RELIGION	Lutheran majority, Roman Catholic and Orthodox minorities
CURRENCY	lat
ORGANIZATIONS	UN

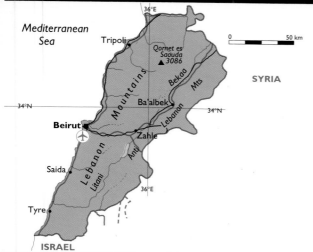

PHYSICAL GEOGRAPHY
Beyond the coastal strip two parallel ranges, known conventionally as the Lebanon and Anti-Lebanon mountains, run the length of the country and are separated by the Bekaa Valley.

CLIMATE
The climate in Beirut is hot in summer, with maximum temperatures exceeding 30°C (86°F), and mild in winter. Rainfall occurs in the winter months, ranging between 920 mm (36 inches) on the coast and 2,300 mm (91 inches) in the mountains.

POLITICS AND RECENT HISTORY
Five years after the Taif Agreement, which ended years of war, invasion, internal trouble and political assassination, the government, led by Prime Minister Rafiq Hariri, is beginning to achieve stability. The Taif Agreement prescribed power sharing, whereby the president is a Maronite Christian, the parliamentary speaker is a Shi'a Muslim and the prime minister a Sunni Muslim. Differences, in particular between the prime minister and the speaker, have prompt-

ed the president to resign on several occasion, but so far he has always been reinstated.

There is no doubt that stability is largely dependent on the powerful influence of Syria, who continues to maintain armed forces in the country. Israel also exerts an influence, as it has not abandoned its sponsorship of the South Lebanon Army, which it maintains as a buffer against the threat of hostility from the North.

ECONOMY
Lebanon is beginning to rebuild itself, and observers claim that economic recovery will continue. As yet the inflow of much needed funds has been slow, as foreign investors wait for assur-

ance that political stability has indeed returned. Confidence is nevertheless growing and the Lebanese pound has remained generally firm.

The government is investing in reconstruction and infrastructure projects, such as the installation of reliable power distribution and telecommunication systems. Similarly, Beirut airport is to be reconstructed.

Although long-term prospects are good, recovery is slow and there have been declines in the value of both agricultural and industrial exports. The Lebanese population, impatient for improved wealth, is beginning to lose patience with the government and further international aid to boost and speed recovery is likely to be necessary. Eventually, however, Beirut may recover its previous status as a regional, financial and cultural centre.

THE REPUBLIC OF LEBANON	
STATUS	Republic
AREA	10,400 sq km (4,015 sq miles)
POPULATION	2,806,000
CAPITAL	Beirut (Beyrouth)
LANGUAGE	Arabic, French, English
RELIGION	62% Shi'a and Sunni Muslim, 38% Roman Catholic and Maronite Christian
CURRENCY	Lebanese pound (LBP)
ORGANIZATIONS	Arab League, UN

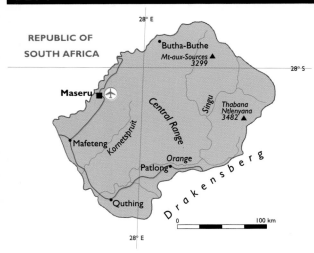

PHYSICAL GEOGRAPHY

Lesotho is completely encircled by South Africa. It is a mountainous country, especially where the Drakensberg mountains rise to heights well over 3,000 m (9,800 ft). From these peaks the land slopes westwards in the form of dissected plateaux.

CLIMATE

Climatic conditions are very much influenced by altitude, to the extent that snowfall is common in the mountains. In general, however, the climate is warm and pleasant with prolonged sunshine. Rainfall, sometimes variable, falls mainly in the summer months.

POLITICS AND RECENT HISTORY

Lesotho achieved independence from its former British protected status as Basutoland in 1966. It is a constitutional monarchy in which King Letsie III acceded to the throne when his father, King Moshoeshoe, was deposed by the ruling military council in 1990. The council had seized power in a coup in 1986 in which Chief Jonathon, prime minister since independence, was replaced.

The military council promised a return to civilian rule, and in general elections in March 1993 the Basutho Congress Party gained a clear victory over the Basutho National Party. Its leader, Dr Ntsu Mokhehle was sworn in as prime minister. The military council kept to their word, but interference from them was still prevalent. Claiming that the government had lost popular support, King Letsie dismissed it in August 1994, but under international pressure, notably from President Mandela of South Africa and President Mugabe of Zimbabwe, was forced to concede power to Dr Mokhehle a month later. In early 1995, King Letsie himself abdicated, allowing the monarchy to be restored to his father King Moshoeshoe. This gives rise to hopes for future political stability.

ECONOMY

Lesotho is a relatively poor country, but the return of civilian rule has engendered a degree of optimism about the kingdom's prospects. The government has introduced economic and financial reforms but, for the time being, the country is dependent upon international aid.

Agriculture, the mainstay of the economy, is subject to the vagaries of climate and is restricted to the lowlands and foothills, because of the terrain. The main crops are maize, sorghum and wheat, but experimental cultivation of sunflowers is promising.

Whatever economic steps the Lesotho government may take, the nation's prosperity is inevitably bound up with political and economic events in South Africa.

THE KINGDOM OF LESOTHO	
STATUS	Kingdom
AREA	30,345 sq km (11,715 sq miles)
POPULATION	1,943,000
CAPITAL	Maseru
LANGUAGE	Sesotho, English
RELIGION	80% Christian
CURRENCY	loti (LSL) South African rand (ZAR)
ORGANIZATIONS	Comm., OAU, SADCC, UN

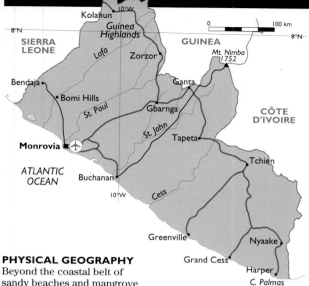

PHYSICAL GEOGRAPHY
Beyond the coastal belt of sandy beaches and mangrove swamps, the land rises to a forested plateau.

CLIMATE
The climate is hot, with average daily temperatures exceeding 25°C (77°F) throughout the year. Rainfall is plentiful and particularly heavy in June and July.

POLITICS AND RECENT HISTORY
Civil war broke out in 1989 when the National Patriotic Front of Liberia (NPFL), under Charles Taylor, invaded from the Ivory Coast, seeking to free the country from the authoritarian rule of President Samuel Doe. Within months Doe had been deposed and assassinated. Neighbouring states, under the aegis of the Economic Community of West African States, anxious about the regional instability caused by the fighting, provided a peace-keeping force known as ECOMOG. A peace agreement was reached in October 1991 and a provisional government installed under Amos Sawyer. However, a year later fierce fighting broke out again, the main factions being the NPFL, the United Liberation Movement and ECOMOG, which supported the provisional government.

In 1993 some hope for peace emerged with agreement to disarm the insurgents and to hold free elections the following year. These hopes proved short-lived. The disarmament did not take place, factional fighting, associated with atrocities, broke out once more and the elections were cancelled.

There is no immediate prospect of peace. Liberia is divided among at least seven war lords. ECOMOG forces, of whom the majority are Nigerian, contribute little to stability and actively support factions opposed to the NPFL.

ECONOMY
War has ravaged the Liberian economy. United Nations relief is necessary, but can not reach most of the population, close to starvation, nearly half of whom have been displaced from their homes by the fighting. The capital Monrovia has been looted by ECOMOG.

Yet, before the civil war, Liberia enjoyed a measure of prosperity. Iron ore deposits provided about 50 per cent of the nation's revenue, but the iron ore refinery has been stripped and shipped to Nigeria. In addition, the country had extensive rubber plantations, mainly foreign owned.

Important revenue also accrued from the registration of a shipping fleet which, at a tonnage exceeding 50 million, is the world's largest. Although this survives, there is no government in place to make effective use of the revenue. As yet, no economic reconstruction is possible and, when it is, massive international involvement will be necessary.

THE REPUBLIC OF LIBERIA	
STATUS	Republic
AREA	111,370 sq km (42,990 sq miles)
POPULATION	2,640,000
CAPITAL	Monrovia
LANGUAGE	English, tribal languages
RELIGION	traditional beliefs, Christian, Muslim
CURRENCY	Liberian dollar (LRD) US dollar (USD)
ORGANIZATIONS	ECOWAS, OAU, UN

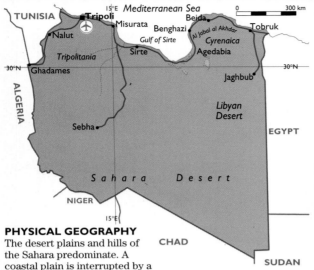

PHYSICAL GEOGRAPHY

The desert plains and hills of the Sahara predominate. A coastal plain is interrupted by a mountain range east of Benghazi.

CLIMATE

The climate is hot and arid with an average rainfall of less than 130 mm (5 inches) over 95 per cent of the country. Only the Mediterranean coast receives winter rainfall sufficient to sustain agriculture, around 200–610 mm (8–24 inches annually.

POLITICS AND RECENT HISTORY

Libya is a dictatorship. Colonel Muammar Gaddafi, designated as Leader of the Revolution, came to power after the coup which deposed King Idris in 1969. Although Colonel Gaddafi does not now hold a formal position in government, he effectively wields unfettered power. Opposition groups are banned in Libya and operate only in exile.

Over the years, Gaddafi has regularly come into conflict with western nations over his alleged sponsorship of terrorism, most recently in connection with the bombing of the Pan Am aircraft over Lockerbie, Scotland in 1988.

His refusal to release two suspects for trial has prompted United Nations trade sanctions. From time to time, signs of unrest can be detected among the Libyan population, but any opposition is ruthlessly suppressed.

Since the Gulf War, Libya has improved its relationship with other Arab nations, notably Egypt, which has given cautious support to Libya's position in respect of the Lockerbie bombing.

ECONOMY

Oil, the source of 80 per cent of gross domestic product, has transformed the Libyan economy from poverty in 1960 to one of comparative wealth.

THE GREAT SOCIALIST PEOPLE'S LIBYAN ARAB JAMAHIRIYA	
STATUS	Republic
AREA	1,759,540 sq km (679,180 sq miles)
POPULATION	4,700,000
CAPITAL	Tripoli (Tarabulus)
LANGUAGE	Arabic, Italian, English
RELIGION	Sunni Muslim
CURRENCY	Libyan dinar (LYD)
ORGANIZATIONS	Arab League, OAU, OPEC, UN

Economic sanctions have so far had little effect as oil is excluded – western nations, especially Italy, Germany and Spain, are heavily dependent on Libyan oil and would be badly affected by a halt in supply. American attempts to apply the sanctions to oil are, therefore, unlikely to be successful. In any case, there is widespread flouting of those sanctions that have been agreed.

Libya is dependent on a large immigrant labour force from its Arab neighbours, and also from some Asian countries. Immigrants are employed in the oil industry and on major construction projects. Of these, the two with the highest priority are a railway traversing the country to Tunisia and Egypt, and the Great Man Made River Project. This ambitious water pipeline project is aimed at bringing irrigation to arid areas. It was initiated in 1984 and the first phase was completed in 1991. However, the final phase, though scheduled for completion in 1997, is likely to be delayed. More immediate progress on the railway seems likely, as Libya and Egypt have signed an agreement for the reconstruction of the Italian railway along the Libyan coast, destroyed in World War II.

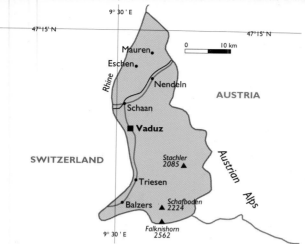

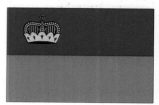

PHYSICAL GEOGRAPHY

Two physical zones are evident – the fertile flood plains of the upper Rhine valley in the north and west and the mountains of the Austrian alps in the east.

CLIMATE

The climate is temperate, with cool winters and snowfall on high ground and warm wet summers.

POLITICS AND RECENT HISTORY

The principality of Liechtenstein is a constitutional monarchy. The monarch is HSH Prince Hans-Adam II, who succeeded as head of state in 1990 following the death in November 1989 of his father HSH Prince Francis Joseph II, who had ruled for 51 years. Women were given the right to vote in 1984. In the 1986 election the Fatherland Party (VU), in power since 1970, retained control. In further elections in 1989 the number of seats in the parliament was increased from 15 to 25 and the VU narrowly won. In elections of February 1993, the Progressive Citizens' Party gained 12 seats to the VU's 11. In September 1993 the prime minister, Markus Buechel, suf-fered a vote of no confidence. Prince Hans-Adam dissolved parliament and in the elections that followed, the VU regained its majority under Prime Minister Mario Frick.

ECONOMY

Before 1945 the principality was primarily agricultural. In 1930 farming engaged 70 per cent of the population, but by 1991 this figure had fallen to 1.7 per cent. Liechtenstein is now highly industrialized, with a variety of light industries, often specialist manufacturing such as precision instruments and dental equipment, pharmaceuticals, ceramics and textiles. The metal working industry, however, employs more than twice as many workers as all the other main industries combined. Industrialization led to immigration from neighbouring European countries and there is now a high proportion of foreign labour.

Other main sources of revenue are postage stamps, food products and tourism. The fastest-growing sector, however, is banking, because of the very low taxation and strict laws on banking secrecy. With such success, it is not surprising that Liechtenstein enjoys the world's highest per capita gross domestic product.

The country has a long-standing customs union with Switzerland, uses Swiss currency and, since 1919, has had an agreement that it may be represented by that country abroad. Despite this association, Liechtenstein, unlike its neighbour, voted at a referendum to join the European Economic Area and was formally admitted on 1 May 1995. The close relationship with Switzerland nevertheless survives.

THE PRINCIPALITY OF LIECHTENSTEIN	
STATUS	Principality
AREA	160 sq km (62 sq miles)
POPULATION	30,000
CAPITAL	Vaduz
LANGUAGE	Alemannish, German
RELIGION	87% Roman Catholic
CURRENCY	franken (Swiss franc) (CHF)
ORGANIZATIONS	Council of Europe, EEA, EFTA, UN

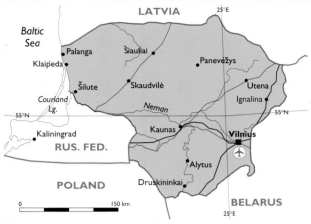

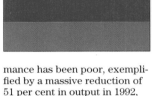

PHYSICAL GEOGRAPHY

Lithuania is low-lying, with many lakes and ridges of glacial origin, often pine covered.

CLIMATE

The climate is essentially temperate in nature, although winter temperatures are decidedly cold. Most rain falls in the summer months.

POLITICS AND RECENT HISTORY

In 1990 the Sajudis Nationalist Party under Vytautas Landsbergis won election victory, declared independence and thereby provoked the intervention of Soviet forces. In September 1991 independence was conceded by the Soviet Union and recognized internationally. Sajudis was not able to maintain its pre-independence support and in October 1992 the Lithuanian Democratic Labour Party (LDLP), whose origins were in the former Communist Party, won an overall majority in general elections. Similarly, in February 1993, their leader Algirdas Brazauskas, who had been the Communist leader, won a comfortable victory in presidential elections, deposing Landsbergis. A month later, President Brazauskas appointed Adolfas Šleževičius as prime minister, who formed a new government

The Lithuanian government has placed a premium on maintaining cordial relations with Russia, and this has proved more successful than in the other Baltic States. Brazauskas was able to reach agreement with President Yeltsin on the early withdrawal of the remaining Russian troops from Lithuania, and by September 1993 the last were repatriated. In January 1995, a new law established Lithuanian as the official state language.

ECONOMY

Lithuania has managed the economic transition from Soviet status to independent nationhood less well than either Estonia or Latvia, even though better relations have been maintained with Russia. Economic performance has been poor, exemplified by a massive reduction of 51 per cent in output in 1992, compared with 1991. This drop, accompanied by rising unemployment, was the decisive factor in the electoral defeats of Sajudis.

The present government, seeking to create a market economy, has established six free economic zones aimed at attracting foreign businesses. However, progress in privatization has been sluggish and land reforms have yet to benefit Lithuanian farmers. Generally, the economy is still firmly linked with the ex-Soviet countries; a mere 10 per cent of the nation's products travel westwards.

The economy is primarily agricultural, with beef and dairy produce dominant. Major crops include potatoes and flax, and there is a large fishing industry. Industrial products include paper, chemicals, electronics and electrical goods.

International concern has focused on the nuclear power generating plant, of the Chernobyl type, at Ignalina. Lithuania cannot afford to shut this down and Sweden is investing capital with the aim of achieving safer operation.

THE REPUBLIC OF LITHUANIA	
STATUS	**Republic**
AREA	**65,200 sq km (25,165 sq miles)**
POPULATION	**3,735,000**
CAPITAL	**Vilnius**
LANGUAGE	**Lithuanian, Russian, Polish**
RELIGION	**80% Roman Catholic**
CURRENCY	**litas**
ORGANIZATIONS	**Council of Europe, UN**

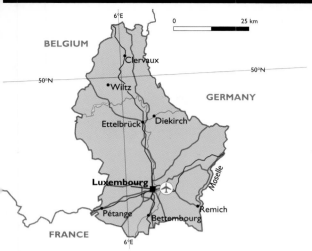

PHYSICAL GEOGRAPHY

The hills and forests of the Ardennes dominate the northern third of the country with rolling pastureland to the south.

CLIMATE

The climate is temperate, with rainfall ranging from 700–1,000 mm (28–40 inches) a year.

POLITICS AND RECENT HISTORY

Despite its relatively small size, the Grand Duchy of Luxembourg plays a prominent role in the affairs of Europe. Luxembourg was a founder member of the North Atlantic Treaty Organization in 1949 and of the European Economic Community in 1957, and was host to the European parliament before 1989. In consequence, the nation exerts an influence that is disproportionate to its size, and is concerned about imminent moves within the European Union (EU) to reduce the authority of smaller nations.

The country enjoys political stability. A coalition led by the Christian Social People's party has ruled since 1984 and, under Prime Minister Jacques Santer, won a further term in elections in mid-1994. On Santer's appointment to President of the European Commission at the beginning of 1995, Jean-Claude Juncker was appointed prime minister.

ECONOMY

Luxembourg enjoys a stable and prosperous economy with low inflation and the lowest rate of unemployment in the EU. Over half the population are foreign workers attracted by the employment opportunities. Trading, mainly with the EU countries, is vital.

Recessions in Germany and particularly in Belgium (to which Luxembourg's currency is linked) have affected the Duchy's own economy. The steel industry, the traditional basis of Luxembourg's wealth, was most affected. Nevertheless, other sectors showed a surge in activity, most notably the banking and finance centre. Luxembourg has become a leading centre in Europe for banking and finance, encouraged by low tax rates and the government's strict secrecy laws. It has launched a bid to host the European Central Bank, and is already home to the European Investment Bank which supports economic integration in Europe.

Overall, evidence of economic growth in the Duchy is exemplified in the new business park near Luxembourg airport which, at one time, was advertising for tenants and is now nearly full.

THE GRAND DUCHY OF LUXEMBOURG	
STATUS	Grand Duchy
AREA	2,585 sq km (998 sq miles)
POPULATION	401,000
CAPITAL	Luxembourg
LANGUAGE	Letzeburgish, French, German, English
RELIGION	95% Roman Catholic
CURRENCY	Luxembourg franc (LUF) Belgian franc (BEF)
ORGANIZATIONS	Council of Europe, EEA, EU, NATO, OECD, UN, WEU

PHYSICAL GEOGRAPHY
Macedonia is a rugged country with the highest land in the Sar range in the northwest. The Vardar valley traverses the country from north to south, creating a major strategic route.

CLIMATE
The climate has fine hot summers but bitterly cold winters.

POLITICS AND RECENT HISTORY
Macedonia was a republic within Yugoslavia until November 1992, when it declared its independence. The country faced severe problems in gaining international recognition because of Greek objections to its name. Greece was concerned because it has a province of the same name and feared that use of the name Macedonia, by the newly independent country, would provoke claims upon Greek territory. Its response was to apply a trade embargo.

Nevertheless, most European nations have recognized the Republic which for the time being exists under the formal title authorized by the United Nations (UN) – the Former Yugoslav Republic of Macedonia.

A dialogue between Greece and Macedonia has begun, under UN supervision. It seems likely that Greece will withdraw its objections and will merely seek assurances that Macedonia will make no claims on Greek territory, and will adopt a flag that omits the symbol used at present.

President Kiro Gligorov, who has managed the succession from Yugoslavia, won a further five-year term in parliamentary and presidential elections at the end of 1994. However, his government needs the support of the Albanian minority who are becoming more vociferous. Ethnic tensions are likely to increase.

FORMER YUGOSLAV REPUBLIC OF MACEDONIA	
STATUS	Republic
AREA	25,715 sq km (9,925 sq miles)
POPULATION	2,083,000
CAPITAL	Skopje
LANGUAGE	Macedonian, Albanian minority
RELIGION	Orthodox
CURRENCY	denar
ORGANIZATIONS	Council of Europe, UN

ECONOMY
The Macedonian economy is in considerable difficulty and has suffered particularly from the UN trade sanctions imposed on Serbia. Seventy per cent of Macedonia's trade was formerly with the remainder of Yugoslavia. The conflict over the republic's name has also damaged trade with and through Greece, the natural link with the outside world. The government has in consequence forged closer trading ties with both Bulgaria and Turkey, whose governments were among the earliest to recognize Macedonia's independence. Its equivocal status and its geographical position in a region of strife and tension have inhibited foreign governments from providing both aid and investment.

As a result, Macedonia faces a massive foreign debt. It suffers from high inflation and average income has fallen dramatically in recent years. The need for foreign assistance is urgent if stability is to be maintained, and may now be forthcoming if the quarrel with Greece is resolved. Even so, prospects are not good. The main crop, tobacco, has too high a tar content to be acceptable in western markets.

PHYSICAL GEOGRAPHY
Madagascar, the world's fourth largest island, is a high plateau with a narrow lowland strip along the east coast.

CLIMATE
The climate is tropical with heavier rainfall to the north and monsoons affecting the east coast, where rainfall reaches 1,500–2,000 mm (59–79 inches) a year. Southern parts of the island are arid.

POLITICS AND RECENT HISTORY
In 1960, Madagascar achieved full independence from its former status as the autonomous Malagasy Republic within the French Community. For 17 years, until 1991, it was a one-party state under President Didier Ratsiraka. His government was undermined by widespread civil unrest following a massacre in August 1991, when the army fired into a crowd during a general strike. Anger caused by this incident was believed to have inspired the demand for democracy. In October 1991, Ratsiraka agreed to a transitional government with opposition leaders which, he would, nevertheless, lead.

In February 1993 Albert Zafy defeated Ratsiraka in presidential elections, becoming the first president of the new Third Republic. The mid-1993 legislative elections were won by Zafy's *Forces Vives* and allied parties, and the parliament elected Francisque Ravony as prime minister. In early 1995, Ravony took over direct responsibility for the country's finances, in order to negotiate with the International Monetary Fund and the World Bank. A new multi-party constitution, effective from 1992, reduced the influence of the president and gave virtually all executive powers to the prime minister.

ECONOMY
Although Madagascar has an essentially agricultural economy, with 85 per cent of the population engaged in farming, cultivation is adversely affected by uncontrolled erosion of the land.

Madagascar has traditionally been the world's leading producer of vanilla, followed by Indonesia and Comoros. Vanilla remains one of the country's major exports, but has shown some decline in recent years owing to competition from artificial vanilla. Coffee, the second major export, offers better prospects as Brazil and Colombia have agreed to limit their production.

In February 1994, Madagascar suffered the worst cyclone since 1927; 150,000 people lost their homes and the main port was destroyed.

The fishing industry has shown significant growth, and prawns are now a valuable export commodity, promising to rival vanilla and coffee in importance. There are also encouraging prospects for other shellfish, such as lobsters, and for tuna. The tourist industry, although undeveloped, shows rapid growth and may provide the key for Madagascar to improve on its position as one of the world's poorest nations.

THE REPUBLIC OF MADAGASCAR	
STATUS	Republic
AREA	594,180 sq km (229,345 sq miles)
POPULATION	12,092,000
CAPITAL	Antananarivo
LANGUAGE	Malagasy, French, English
RELIGION	47% animist, 48% Christian, 2% Muslim
CURRENCY	Malagasy franc (MGF)
ORGANIZATIONS	OAU, UN

PHYSICAL GEOGRAPHY

Malawi is a small hilly country at the southern end of the East African Rift Valley. The dominant feature is Lake Malawi.

CLIMATE

The climate is mainly subtropical, with varying rainfall and a dry season from May to October.

POLITICS AND RECENT HISTORY

From independence in 1964, President Hastings Kamuzu Banda ruled Malawi (formerly Nyasaland), leading the Malawi Congress Party. Although in the 1980s the country became known for its farming achievements, latterly it has become notorious for its human rights abuses. International protests from Britain, the USA and others took the form of sharp cuts in foreign aid, which rapidly had an adverse effect on the economy. This, and growing domestic pressure for reform, forced Dr Banda to hold a referendum in June 1993 to decide between the existing one-party rule and a multi-party system. The result was 63 per cent in favour of democracy. The government was obliged to include the opposition in a National Consultative Council to oversee the transition from single-party rule to multi-party elections in 1994.

In October 1993, while Dr Banda underwent brain surgery, effective power was temporarily transferred to his notorious associate John Tembo. Two months of civil strife ensued, until Dr Banda resumed the presidency in December.

Eventually, in May 1994, the long awaited multi-party elections were held. Dr Banda was defeated and forced to concede power. The United Democratic Front gained most seats, and their leader Bakili Muluzi became president. They did not, however, gain an overall majority and have formed a coalition with the Alliance for Democracy. Shortly after gaining office, the government commissioned an enquiry into the deaths of former government ministers. In response to the enquiry's findings, both Dr Banda and John Tembo have been charged with murder.

THE REPUBLIC OF MALAWI	
STATUS	Republic
AREA	94,080 sq km (36,315 sq miles)
POPULATION	9,135,000
CAPITAL	Lilongwe
LANGUAGE	Chichewa, English
RELIGION	majority traditional beliefs, 10% Roman Catholic, 10% Protestant
CURRENCY	kwacha (MWK)
ORGANIZATIONS	Comm., OAU, SADC, UN

ECONOMY

The economy is predominantly agricultural – 96 per cent of the population work on the land. Maize is the main subsistence crop and the major exports are tobacco, tea and sugar. However, with recent poor harvests due to severe drought, exports have declined.

Malawi has deposits of both coal and bauxite, but these are under-exploited at present. Manufacturing industry concentrates on consumer goods and construction materials. All energy is produced by hydroelectric power.

There is a traditional migration of labour, mainly miners, to other countries, particularly South Africa. Per capita income is among the lowest in the world.

Malawi is heavily reliant on foreign aid, which was curtailed when human rights abuses were rife. Following political reform, future assistance is more assured.

PHYSICAL GEOGRAPHY

The Federation of Malaysia consists of two separate parts. Western Malaysia forms a peninsula with mountain ranges aligned along its axis. Eastern Malaysia, consisting of Sabah and Sarawak on the island of Borneo, is mainly jungle-covered hills and mountains with mangrove swamps along the coast.

CLIMATE

The climate is warm with heavy rainfall, averaging 2,500 mm (98 inches) a year, brought by the prevailing monsoon.

POLITICS AND RECENT HISTORY

The Federation of Malaysia (which included Singapore until its secession in 1965) became independent from the UK in 1963. It is a constitutional monarchy with a king, elected from among the nine state sultans, serving for a five-year term. Executive power is currently held by a government formed by the *Barisan Nasional* (National Front) coalition. The leading party in the coalition is the United Malays National Organization, led by Prime Minister Dr Mahathir Mohamad. At national elections in April 1995, the ruling coalition extended its grip on power, when it gained victory in 162 out of 192 seats, with 63 per cent of the vote. A key policy of the government is to encourage greater participation by the majority Malay population in the nation's economic life. Traditionally it has been the minority Chinese who have wielded economic power.

ECONOMY

Political stability and wise leadership have transformed the Malaysian economy from one which was, in the 1960s, excessively dependent on tin and rubber, to one based on manufacturing. That sector now provides 70 per cent of the nation's exports. Growth has averaged 8 per cent per annum over the past seven years. Industrial activity has grown throughout west Malaysia, with complexes in Johore and on Pinang Island specializing in high-tech products.

Infrastructure has struggled to keep pace with these developments, but the opening, in 1994, of a highway traversing west Malaysia from north to south will assist. Power supplies are often inadequate and the massive Pergau Hydro Electricity project, financed with British aid, has received severe criticism as being an inappropriate solution. The huge aid package also created controversy in Britain following suggestions that the provision of aid was linked to an arms sale. The Malaysian prime minister stated that no further contracts would be given to British companies, although the ban was eventually lifted. Another large hydro-electric project is planned in Sarawak. This will involve the clearance of vast tracts of tropical rainforest and displacement of native peoples.

In east Malaysia, industrial development lags far behind that in the western province, although the island of Labuan is being established as an offshore financial centre.

Although the Malaysian government clearly sees the country's future in manufacturing industry, rather than primary products, Malaysia is still the world's largest producer of timber and timber products. This has attracted widespread criticism from environmentalists.

THE FEDERATION OF MALAYSIA	
STATUS	Federation
AREA	332,965 sq km (128,525 sq miles)
POPULATION	19,047,000
CAPITAL	Kuala Lumpur
LANGUAGE	58% Bahasa Malaysian, English, Chinese
RELIGION	53% Muslim, 25% Buddhist. Hindu, Christian and animist minorities
CURRENCY	Malaysian dollar or ringgit (MYR)
ORGANIZATIONS	ASEAN, Col. Plan, Comm., UN

MALDIVES

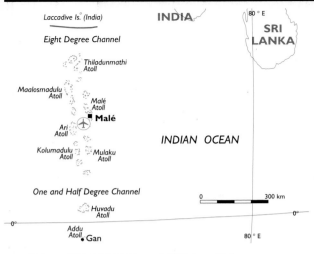

PHYSICAL GEOGRAPHY

The Maldive Islands are a string of well over 1,000 coral atolls, of which about 200 are inhabited. They have developed on submarine volcanoes and the larger ones are covered with lush tropical vegetation.

CLIMATE

The climate is hot throughout the year, with maximum daily temperatures consistently above 30°C (86°F). Humidity is generally high, while the heaviest monsoon rains fall between May and December.

POLITICS AND RECENT HISTORY

The Maldives achieved independence from British protected status in 1965, firstly as a sultanate and three years later as a republic. The president is elected for a five-year term by the people, and he in turn appoints a cabinet which is responsible to the parliament. There are no political parties. President Maumoon Gayoom became president in 1978 when his predecessor decided not to stand, and was re-elected in 1983, 1988 and in late 1993. In 1988 Gayoom survived a coup, with the assistance of Indian troops, inspired by exiled

Maldivians. This event encouraged the president to lead a campaign for better security of states so small that maintenance of an army is not feasible.

ECONOMY

The main source of revenue is tourism, which expanded rapidly during the 1980s following the opening of Malé International Airport in 1981. Tourist arrivals increased from 42,000 in 1980 to 241,000 in 1993, and it is on the further development of this sector of the economy that the future prosperity of the country depends. The government has, therefore, invested in projects such as desalination plants, generators and air conditioning, that are likely to help boost tourism.

Fishing, once the most important economic activity, remains significant. Tuna is exported but is constrained by lack of refrigeration facilities, although the fishing fleet is now mechanized. The Maldives also maintain a merchant shipping fleet, but revenue has declined as this industry suffers worldwide recession.

There is little cultivable land but bananas, sweet potatoes, mangoes and coconuts, which provide some export revenue, are grown. Most food stuffs are nevertheless imported

The islands are all low-lying and are threatened by marine incursions, although sea defenses are being improved with Japanese financial assistance.

The Maldives has been classified as one of the world's poorest countries but, with gross domestic product growth averaging 10 per cent per annum during the past 20 years and with the development of the tourist trade, the future looks promising.

THE REPUBLIC OF MALDIVES	
STATUS	Republic
AREA	298 sq km (115 sq miles)
POPULATION	238,000
CAPITAL	Malé
LANGUAGE	Dhivehi
RELIGION	Sunni Muslim majority
CURRENCY	rufiyaa (MVR)
ORGANIZATIONS	Col. Plan, Comm., UN

PHYSICAL GEOGRAPHY

Northern regions lie within the Sahara desert. To the south the land, through which the River Niger flows, is savannah.

CLIMATE

The climate is hot, with average temperatures for most of the year approaching 30°C (86°F). Rainfall, virtually absent in the north, usually occurs in appreciable quantities during summer in the south, but droughts are increasingly common.

POLITICS AND RECENT HISTORY

In April 1992, the first free elections since independence from France in 1960 were held. In the 1992 elections Alpha Konaré secured the presidency, and with it executive power. He appointed Younoussi Touré as prime minister, but Touré's tenure ended in 1993 following rioting, when he was replaced by Abdoulaye Sow, who in turn has been replaced by Ibrahim Boubacar Keita. A fragile coalition government is dominated by a partnership of the Alliance for Democracy in Mali and Tuareg

representation, the National Pact. The main opposition is the National Committee for Democratic Initiative (CNID), but in March 1995 tension within the CNID caused a split in the party.

Successive governments have been faced with rebellion by Tuareg nomadic peoples in the north of the country. However, in May 1993, the conflict was declared to be over following the peace-making efforts of Tuareg leader Rhissa ag Sidi Mohamed and President Konaré, with the mediation of the Burkinan government. Following the declaration, a two-year project was announced for the repatriation of 100,000 Tuareg refugees, mainly from southern Algeria. Although the rebellion is officially over, there still remains continued tension accompanying the reintegration of Tuareg fighters.

THE REPUBLIC OF MALI	
STATUS	Republic
AREA	1,240,140 sq km (478,695 sq miles)
POPULATION	10,135,000
CAPITAL	Bamako
LANGUAGE	French, Bambara
RELIGION	65% Muslim, 30% traditional beliefs, 1% Christian
CURRENCY	CFA franc (W Africa) (XOF)
ORGANIZATIONS	ECOWAS, OAU, UN

ECONOMY

Mali remains a very poor country, despite improvements in agriculture following recent good harvests and a programme of economic liberalization. Less than 2 per cent of the land can be cultivated. Cotton and groundnuts are the cash crops, otherwise farming is at a subsistence level, characterized by livestock rearing. Severe droughts in recent years have resulted in heavy losses of livestock, which is the basis of the Tuareg's livelihood.

In southwest Mali a new gold mine is expected to come into operation in 1997, and the Kalana gold mine in southern Mali is being reopened.

Revenue is diminished by a long-standing and widespread problem of smuggling across the country's lengthy borders. Small manufacturing industries must compete with neighbouring countries, Nigeria in particular. President Konaré has sought aid from the European Union to consolidate the democratic process in his country. Such aid will doubtless be necessary if the country is to rise from its poverty-stricken status.

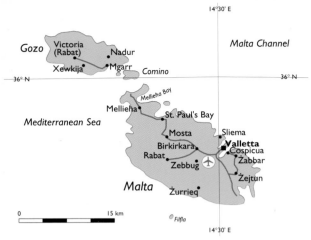

PHYSICAL GEOGRAPHY
The islands of Malta have assumed historical importance for their strategic position in the Mediterranean and their deep natural harbours. The main island of Malta has a landscape of low hills and an indented coastline with numerous beaches and coves. Two-thirds of the population live in the Valletta region.

CLIMATE
The climate is Mediterranean, with warm dry summers averaging 23°C (73°F) and mild winters in the region of 14°C (57°F). Rainfall, on average 500 mm (20 inches) a year, falls mainly between October and March.

POLITICS AND RECENT HISTORY
Malta became independent from the UK in 1964, and a republic in 1974. The National Party won power in 1987, after 16 years of labour domination. Led by Prime Minister Dr Edward Fenech-Adami, it retained control in the last elections, held in 1992.

The major political issue confronting Malta has been the possibility of membership of the European Union (EU).

Malta applied for membership in 1990 and was informed that it would require extensive structural reform. It would, for example, need to reduce subsidies to the shipbuilding and repair industry, in order to comply with the EU's competition rules. Malta's principles of neutrality might also lead to difficulties in any future co-operation on defence. The Maltese government set about liberalizing the economy and by early 1995 had earned praise for the steps taken so far, in a European Commission report . The probable result is that negotiations on entry will start in earnest in 1997.

ECONOMY
When the British naval base closed in 1979, Malta had to resort to diversification in its

THE REPUBLIC OF MALTA	
STATUS	Republic
AREA	316 sq km (122 sq miles)
POPULATION	366,000
CAPITAL	Valletta
LANGUAGE	Maltese, English, Italian
RELIGION	Roman Catholic majority
CURRENCY	Maltese lira (MTL)
ORGANIZATIONS	Comm., Council of Europe, UN

economy. The shipbuilding and repair industries, traditional in Malta, became more important In the absence of any natural resources, light industries were developed based on imports of raw materials, for instance clothing, shoes, plastics and electronics. These industries were intended to boost exports, the main trading partners being Italy, Germany and the UK, and new legislation in 1988 encouraged foreign investment.

National revenue comes chiefly from the service industries, which provide 43 per cent of employment, together with a growing tourist trade, with about one million visitors annually. Malta is self-sufficient in agricultural products and some, such as grapes, tomatoes, onions and potatoes, are exported.

At present, Malta runs a balance of payments deficit because of the need to import key commodities such as oil. However, the development of a large container port at Marsaxlokk may eventually reverse this situation. This project envisages Malta as a hub for trans-shipment to and from many centres in Europe.

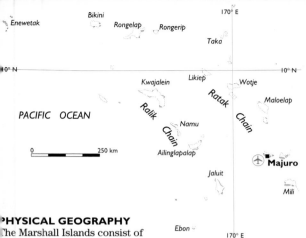

PHYSICAL GEOGRAPHY

The Marshall Islands consist of over 1,000 atolls, islands and islets, generally within two chains, the Ratak (Sunrise) and Ralik (Sunset), but which together account for only 181 sq km (70 sq miles) of land area. The principal atolls are Majuro, Kwajalein, Jaluit, Enewetak and Bikini.

CLIMATE

The climate is tropical and hot, with temperatures near 27°C (80°F) all the year round, but it is cooled by sea breezes and, for three months from December, by the trade winds. Rainfall is heavy, averaging 4,050 mm (160 inches) annually, falling mainly between August and November.

POLITICS AND RECENT HISTORY

At the end of World War II, the Marshall Islands became part of the United Nations (UN) Trust Territory of the Pacific Islands and were administered by the United States. In 1986 sovereign self-governing status was granted in a Compact of Free Association with the USA, in which the latter terminated its administration but remained responsible for defence and security. Sovereign status was not recognized by the UN until December 1990, when the UN Trusteeship officially came to an end, and on 17 September 1991 the Marshall Islands were admitted to UN membership.

The head of state, President Amata Kabua, was elected in 1979 and re-elected for the third time in 1992. The legislature is a 33-member parliament, the *Nitijela*.

The Marshall Islands are a member of the South Pacific Forum, and one of the issues raised in recent years has been the threat of a rising sea-level following global warming. UN studies indicate that a rise of only several metres would submerge the territory. Contingency plans are being devised for evacuation of the Marshall Islanders and others in the region to Australia and New Zealand.

ECONOMY

The islands are lacking in natural resources, apart from some phosphate deposits mined on the atoll of Ailinglapalap. The population is engaged in fishing and subsistence farming, although some products are available for export such as coconuts, melons and tomatoes. The tourist trade is centred on Majuro, which has over 130 hotels and is a base for inter-island cruises.

The economy of the Marshall Islands is predominantly dependent on grants from the USA for the use of the islands as military bases. The US government has recently increased its use of the existing base on Kwajalein Atoll.

The Marshall Islands include Bikini atoll, famous for nuclear bomb testing in the 1950s and uninhabitable ever since. The government is pursing the possibility of developing a nuclear waste capability on one of the islands of the atoll, but has not won the support of either Washington or Tokyo.

THE REPUBLIC OF THE MARSHALL ISLANDS	
STATUS	Self-governing state in Compact of Free Association with USA
AREA	605 sq km (234 sq miles)
POPULATION	45,563
CAPITAL	Majuro
LANGUAGE	English, local languages
RELIGION	Roman Catholic majority
CURRENCY	US dollar (USD)
ORGANIZATIONS	UN

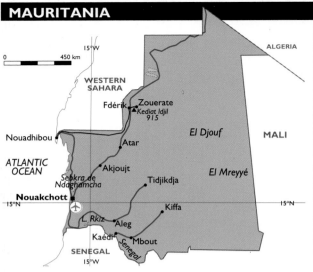

PHYSICAL GEOGRAPHY

Almost the entire area of Mauritania is desert, much of it completely uninhabitable. The only area of significant permanent vegetation is along the River Senegal at the country's southern border. Generally, the terrain is flat plateau which nowhere exceeds 1,000 m (3,300 ft).

CLIMATE

Summers are extremely hot and winters are mild. Inland areas receive little or no rain. Along the coast, light rain falls in summer.

POLITICS AND RECENT HISTORY

Mauritania achieved independence from France in 1960. Subsequently, when Spain withdrew from Western Sahara, Mauritania agreed on a division of that country with Morocco. However, the government was unable to withstand the activities of Polisario guerrillas and renounced its claim in 1979. By then a military government was installed, which later imposed single-party rule under the Military Committee for National Salvation.

In 1991 a referendum produced a massive vote in favour of multi-party democracy. This was followed by the first presidential elections, with several contenders for office. The incumbent, President Taya, won comfortably over his nearest rival, Ahmed Ould Daddah. This victory brought complaints of vote rigging, possibly justified, but the result has stood. The presidential elections were followed a few weeks later by elections to the National Assembly. President Taya's Democratic and Socialist Republican Party won a sweeping victory, but leading opposition groups had boycotted the election.

ECONOMY

The economic prosperity of Mauritania is closely dependent upon climatic conditions. Traditional cattle rearing has been seriously undermined by drought. Year by year the scant rains diminish, the Sahara desert advances and grazing land is destroyed. This has caused major population migrations to towns such as Nouakchott and Kiffa, in turn causing social problems and in particular food shortages, forcing the government to carry out distribution of grain and dairy produce. The introduction of value added tax in 1995, which meant huge rises in the price of food, provoked rioting in Nouakchott.

Arable farming is practised in the south of the country, where the waters of the Senegal River are sufficient to allow irrigation. The Senegal River Project promises to extend cultivable land in the area. At present the typical crops are millet, rice and beans, but eventually cash crops of cotton and sugar cane could be grown.

In the meantime the nation is dependent upon its mineral reserves. The most important of these is iron ore in the northwest at Fdérik, which is connected by rail to the port of Nouadhibou. Mauritania also has substantial copper reserves at Akjoujt.

THE ISLAMIC REPUBLIC OF MAURITANIA	
STATUS	Islamic Republic
AREA	1,030,700 sq km (397,850 sq miles)
POPULATION	2,161,000
CAPITAL	Nouakchott
LANGUAGE	Arabic, French
RELIGION	Muslim
CURRENCY	ouguiya (MRO)
ORGANIZATIONS	Arab League, ECOWAS, OAU, UN

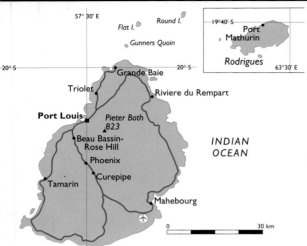

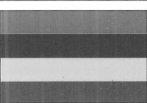

PHYSICAL GEOGRAPHY

Mauritius comprises a main island and about 20 smaller ones. The main island has a coral coast rising to a fertile central plateau.

CLIMATE

The climate is warm with tropical storms, mostly between December and March. The central plateau receives considerably more rain, averaging 5,100 mm (200 inches), than the 1,000 mm (40 inches) experienced elsewhere.

POLITICS AND RECENT HISTORY

Mauritius became independent from the UK and joined the Commonwealth in 1968. In March 1992 the state became a republic and Governor-General Sir V. Ringadoo became president until Cassam Uteem's nomination in June 1992. In 1990, Sir Anerood Jugnauth, who has been prime minister since 1982, forced an alliance of two parties, the *Mouvement Militant Mauricien* and the *Mouvement Socialiste Militant*. This combination defeated the main opposition Labour party in the 1991 general election, but there was some doubt whether Sir Anerood can maintain power at the next elections.

ECONOMY

Ten years ago Mauritius had practically a monocultural economy based on sugar. Inflation was over 30 per cent and unemployment was at 20 per cent. Since then, the economy has shown sustained growth of nearly 6 per cent a year, there is virtually no unemployment and the population, which at one time was growing rapidly, is now stable.

Although sugar is still a vitally important crop, and occupies 90 per cent of cultivable land, the economy is not totally dependent on it. Clothing manufacture and tourism are also important revenue earners. The value of the sugar crop, despite being particularly suited to Mauritius in that it is resistant to cyclones, is set to decline following the last General Agreement on Tariffs and Trade round of talks. Some diversification into other crops, such as tea, has already started.

The textile industry has grown following the development of an export processing zone. This industry, mainly knitwear clothing, generates revenue which now exceeds that of sugar.

The tourist industry has expanded dramatically in recent years. The country's dilemma is how much further the industry should grow before it risks damaging the island.

Further growth has been registered in the finance sector, based on financial services, offshore banking and free port facilities.

THE REPUBLIC OF MAURITIUS	
STATUS	Republic
AREA	1,865 sq km (720 sq miles)
POPULATION	1,098,000
CAPITAL	Port Louis
LANGUAGE	English, French, Creole, Hindi
RELIGION	51% Hindu, 31% Christian, 17% Muslim
CURRENCY	Mauritian rupee (MUR)
ORGANIZATIONS	Comm., OAU, SADC, UN

PHYSICAL GEOGRAPHY
The greater part of Mexico is high plateaux flanked by the west and east Sierra Madre mountain ranges. Altitudes increase towards the south with peaks well over 5,000 m (16,500 ft). The greatest extent of lowland is the limestone Yucatan peninsula.

CLIMATE
The climate varies with latitude and altitude. The north is arid but heavy rainfall prevails in the tropical far south. The central plateaux are mild temperate, but with sharp diurnal variations, while the coastal plains are hot and humid.

POLITICS AND RECENT HISTORY
Mexico has a federal government under a president in whom power is centralized. The ruling party, the Institutional Revolutionary Party (PRI), has held power since 1929. In March 1994, the PRI's candidate for the August presidential elections, Luis Donaldo Colosio, was assassinated. His replacement, Ernesto Zedillo, was victorious and he took over from the incumbent Carlos Salinas. He has promised judicial and democra-

tic reform and has appointed a cabinet of youthful reformers. In pursuit of these policies, the brother of the former president has been arrested for the murder of Luis Colosio and the former president has himself left the country, amid allegations of corruption and drug dealing in his administration.

For several years, the Mexican government has had to contend with rebellion in Chiapas Province. However, in April 1995 a cease-fire was agreed with the rebels, the Zapatistas.

ECONOMY
When President Salinas came to power in 1988, he introduced reforms that have dramatically transformed the economy. In earlier years economic activity had been con-

centrated on the oil industry, and 70 per cent of manufacturing was devoted to petroleum products. The collapse of oil prices in the 1980s was the catalyst for change. Salinas's reforms included the removal of trade barriers, privatization on a scale unparalleled elsewhere in the world, removal of subsidies and diversification of the manufacturing industry. These measures led to a dramatic fall in the rate of inflation, reduced unemployment and national debt and achieved reasonable growth. As a result, Mexico was able to join with the USA and Canada in the North American Free Trade Agreement at the end of 1993.

However, the peso was overvalued and shortly after Zedillo's appointment a currency crisis developed. The government was forced to let the peso float and then devalue it. By early 1995, it had fallen much further, and although this gave a massive boost to Mexican exports, stringent measures are now proposed in order to put the internal economy back on track.

THE UNITED STATES OF MEXICO	
STATUS	Federal Republic
AREA	1,972,545 sq km (761,400 sq miles)
POPULATION	87,341,000
CAPITAL	Mexico City
LANGUAGE	Spanish
RELIGION	96% Roman Catholic
CURRENCY	Mexican peso (MXP)
ORGANIZATIONS	NAFTA, OAS, OECD, UN

PHYSICAL GEOGRAPHY

Micronesia comprises 607 atolls and islands. A third of the population lives on Pohnpei.

CLIMATE

The climate is hot, temperatures averaging 26°C (79°F) all year round. Humidity is high and rainfall averages 4,930 mm (194 inches) annually.

POLITICS AND RECENT HISTORY

In 1947 Micronesia became part of the US-administered UN Trust Territory of the Pacific. Self-government was established in 1979. In 1986 Micronesia was granted sovereign status in a Compact of Free Association with the USA, the USA retaining responsibility for defence. In 1991 Micronesia became a full member of the UN.

The islands are grouped into four states, each having its own government. There is a national congress and a federal president. Micronesia is a member of the South Pacific Forum which meets to discuss problems in the region.

ECONOMY

Apart from phosphate deposits, there are no natural resources. Most of the population derives a living from fishing or subsistence farming. Phosphates and copra are exported. Other income is from a growing tourist industry and from Japanese, US and Korean fishing fleets which operate within the territorial waters. Economic development is restricted due to the immense distances to markets, the threat of cyclones and an undeveloped infrastructure. Micronesia depends heavily on the financial support of the USA.

THE FEDERATED STATES OF MICRONESIA	
STATUS	Self-governing Federation in Compact of Free Association with USA
AREA	702 sq km (271 sq miles)
POPULATION	105,000
CAPITAL	Palikir
LANGUAGE	English, eight indigenous languages
RELIGION	Christian majority
CURRENCY	US dollar (USD)
ORGANIZATIONS	UN

PHYSICAL GEOGRAPHY

Moldova consists of hilly plains drained by the Prut and Dniester rivers.

CLIMATE

The climate is warm, with cold spells in winter.

POLITICS AND RECENT HISTORY

Much of Moldova, known as Bessarabia, was part of Romania, incorporated into the Soviet Union in 1939. Moldova became independent in 1991, following the break up of the Soviet Union, and joined the Commonwealth of Independent States.

Although the republic has rejected reunification with Romania, the northeastern Trans-Dniester region, originally part of Ukraine and with a mainly Russian population, has declared independence.

ECONOMY

The mainstays of the economy are viticulture, fruit and vegetables which, together with industries such as food processing, account for over 50 per cent of national income.

Immediately following independence, Moldova suffered economic turbulence, but strict control is beginning to show success. Inflation is under control and the new currency is relatively stable. Privatization is taking place, foreign investment is encouraged and the government has agreed future programmes with the International Monetary Fund, to enhance the economy.

Moldova lacks fuel reserves and depends on Russia for crude oil. It is dependent on the Ukraine for access to the Black Sea, and the government is planning to build a port for freight on the River Prut.

THE REPUBLIC OF MOLDOVA	
STATUS	Republic
AREA	33,700 sq km (13,010 sq miles)
POPULATION	4,356,000
CAPITAL	Kishinev
LANGUAGE	Moldovan, Russian, Romanian
RELIGION	Orthodox
CURRENCY	leu
ORGANIZATIONS	CIS, UN

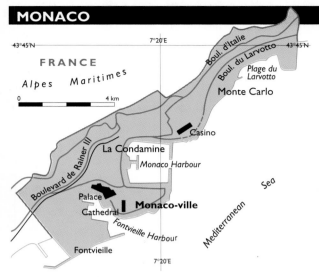

PHYSICAL GEOGRAPHY

Monaco is the world's smallest independent state after the Vatican City. It is situated on a rocky peninsula and occupies a narrow strip of attractive coastline, backed by the foothills of the Alpes Maritimes.

CLIMATE

A mild climate of the Mediterranean variety prevails all the year round. July and August are the hottest months, while January and February are the coolest. Rainfall, generally fairly light, occurs mostly during the winter.

POLITICS

Monaco is a principality ruled by the House of Grimaldi. The present incumbent, Prince Rainier III, has been Monaco's head of state since 1949. Although an independent state, Monaco is formally under French protection and it is the French government that looks after foreign and defence issues. Similarly, Monegasques are treated as if they were French citizens.

In 1962 Prince Rainier drew up a new constitution, following principles of greater democracy, whereby legislative authority is vested jointly in the Prince and an 18-member national council which is elected every five years. Only the National Council can now modify the constitution.

At recent elections, the National and Democratic Party won decisive victories, rendering the liberal and Communist parties virtually ineffective. There has been speculation that Prince Rainier may abdicate, because of failing health, in favour of his son Prince Albert. However, such rumours have been denied by the royal household.

ECONOMY

Since 1958 the principality has increased in size by about 20 per cent due to land reclamation. That land has been assigned mainly for commercial development. As a result, Monaco has benefited from the expansion of light industry, and the present zone is to be reorganized and redeveloped with a view to increasing this sector of the economy. The main products, chemicals, plastics and electronics, form the basis of most external trade with France.

Apart from these industrial activities, the bulk of Monaco's income is derived from the service industries – banking, finance and insurance are particularly significant.

Tourism, important through much of the twentieth century, has become the most profitable source of revenue for the principality. The world-famous casinos contribute a substantial proportion of income.

In recent years, property investment has also brought in wealth and continues to thrive, despite the effects of the recession in the early 1990s.

THE PRINCIPALITY OF MONACO	
STATUS	Principality
AREA	1.6 sq km (0.6 sq miles)
POPULATION	31,000
CAPITAL	Monaco-ville
LANGUAGE	French, Monegasque, Italian, English
RELIGION	90% Roman Catholic
CURRENCY	French franc (FRF)
ORGANIZATIONS	UN

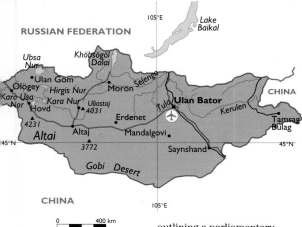

PHYSICAL GEOGRAPHY

Mongolia, three times the size of France, is the world's largest land-locked country. It is characterized by high steppe lands, becoming mountainous with many lakes and rivers in the north and west. In the south is the Gobi desert.

CLIMATE

The Mongolian climate is continental with mild summers and dry, long and very cold winters, prone to heavy snowfalls, when temperatures can fall to -34°C (-29°F). To the south, rainfall averages only 100–130 mm (4–5 inches) a year.

POLITICS AND RECENT HISTORY

Mongolia, as the Mongolian People's Republic, was ruled from 1925 to 1990 as a one-party Communist state, supported by the former Soviet Union. In 1990, facing political pressure, the Communist Party, officially the Mongolian People's Revolutionary Party (MPRP), resigned and political opposition was legalized. In multi-party elections held in 1990 and 1992 the MPRP, headed by Punsalmagiyn Ochirbat, won by a large majority. A new constitution in February 1992,

outlining a parliamentary democracy, redesignated the country the State of Mongolia.

Just before presidential elections in June 1993 President Ochirbat, a strong advocate of reform, was spurned by the MPRP. He was then taken up as leader of a newly-formed opposition, and was re-elected in an easy victory. This result has encouraged the opposition, which is looking forward to the next parliamentary elections in 1996. The MPRP currently dominates parliament, holding 70 out of the 76 seats.

ECONOMY

Mongolia's economy under Communism was closely linked to the Soviet Union, trading meat and raw materials in return for aid. The demise of the Soviet Union, with a consequential loss of aid, and a

breakdown of economic relations with the Commonwealth of Independent States forced Mongolia to make drastic economic reforms. These included a crash programme of privatization of 80 per cent of the economy, facilitated by a government voucher-converting scheme.

Since 1990, however, the economy has been in desperate straits. Exports have been halved, gross domestic product has fallen, inflation is high and many Mongolians live in poverty. A poor wheat harvest in 1994 created the need for food aid.

The economy is based on pastoral activity, but there are rich mineral resources, including coal, copper, tin, gold, lead and tungsten, and half the country's exports come from the Erdenet copper mine. Exploitation of these reserves is hampered by a lack of cash, power supplies are breaking down and equipment needs replacing. However, an Anglo-German consortium is modernizing the Ulan Bator airport.

THE STATE OF MONGOLIA	
STATUS	Republic
AREA	1,565,000 sq km (604,090 sq miles)
POPULATION	2,318,000
CAPITAL	Ulan Bator (Ulaanbaatar)
LANGUAGE	Khalkha Mongolian
RELIGION	some Buddhist Lamaism
CURRENCY	tugrik (MNT)
ORGANIZATIONS	UN

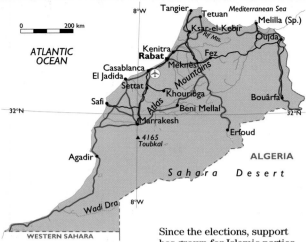

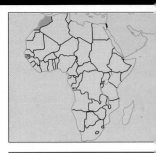

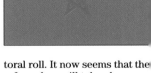

PHYSICAL GEOGRAPHY

The coastal plain is interrupted in the east by the Rif mountains. To the south rise the Atlas mountains, reaching heights of over 4,000 m (13,100 ft), beyond which lie the arid plains of the Saharan fringe.

CLIMATE

Most of Morocco is hot and arid, but the Mediterranean coast experiences more temperate conditions with some winter rainfall.

POLITICS AND RECENT HISTORY

Morocco gained its independence from French and Spanish protectorate status in 1956. It is a monarchy in which King Hassan II exerts substantial authority, with a measure of democracy. Parliamentary elections, the first since 1984, were held in June 1993. No Islamic parties are permitted, and only two-thirds of the parliamentary seats were contested, the remainder being reserved for regional councils, workers' groups and professional associations. Although corruption was evident, notably in the sale of voting cards, the elections were deemed to be reasonably fair.

Since the elections, support has grown for Islamic parties, but there are no signs of the violence associated with fundamentalist groups in other Arab states.

Morocco's annexation of Western Sahara is looking increasingly like a *fait accompli*, although negotiations with the Polisario Front drag on. A cease-fire was declared in 1991, after 16 years of war in the territory. A referendum under United Nations supervision is to be held on the territory's future. There have been many postponements, with Morocco claiming that it has not been possible to compile an electoral roll. It now seems that the referendum will take place, Morocco believing that the outcome will be in its favour.

ECONOMY

The Moroccan economy has been subject to considerable liberalization. State-owned enterprises are being privatized, and foreign investment is encouraged. Inflation is down, and the budget deficit has been reduced, but unemployment is exacerbated by a rapidly growing population and remains high. The government needs to distribute wealth more evenly, if it is to forestall social unrest.

Agriculture is important to the economy, although subject to periodic droughts. For this reason, irrigation projects always carry high priority. The true value of agriculture is difficult to determine, because of the prevalent illegal growing and export of cannabis.

Morocco has some aspirations to join the European Union (EU). However, it has recently been in dispute with the EU, Spain and Portugal. Fishing concessions have not been renewed because Morocco claims that Spain and Portugal have previously exceeded their quotas.

THE KINGDOM OF MOROCCO	
STATUS	Kingdom
AREA	710,895 sq km (274,414 sq miles)
POPULATION	26,069,000
CAPITAL	Rabat
LANGUAGE	Arabic, French, Spanish, Berber
RELIGION	Muslim majority, Christian and Jewish minorities
CURRENCY	Moroccan dirham (MAD)
ORGANIZATIONS	Arab League, UN

WESTERN SAHARA	
STATUS	Territory in dispute, administered by Morocco
AREA	266,000 sq km (102,675 sq miles)
POPULATION	250,000

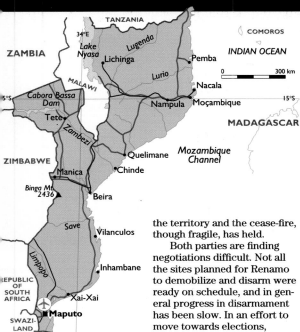

PHYSICAL GEOGRAPHY

The greater part of Mozambique is a plateau with savannah vegetation. Elevations are higher to the north of the country. The coast is fringed by coral reefs and lagoons.

CLIMATE

The climate, hottest along the coast, is tropical, with most rain falling between October and March.

POLITICS AND RECENT HISTORY

Mozambique, formerly an overseas province of Portugal, achieved independence in 1975. Many years of civil war came to an end in October 1992, when a treaty was signed between the Frelimo government of President Joaquim Chissano and the Renamo guerrilla leader Afonso Dhlakama. To support the peace, the United Nations (UN) dispatched 6,000 personnel to the territory and the cease-fire, though fragile, has held.

Both parties are finding negotiations difficult. Not all the sites planned for Renamo to demobilize and disarm were ready on schedule, and in general progress in disarmament has been slow. In an effort to move towards elections, Renamo has dropped its demand for three provincial governorships, while Frelimo has agreed to include ex-rebels in provincial administrations. Elections originally planned for 1993 were not held until October 1994, although some observers thought this date ambitious. Frelimo were victorious and President Chissano retained the presidency. Dhlakama accepted defeat. The UN has now left the country, hopeful that the fledgling democracy will survive. In late 1995 Mozambique opted to join the Commonwealth, the first country not a former UK territory to do so.

ECONOMY

Ravaged by civil war for so many years, Mozambique is an extremely poor country. The economy will require many years of reconstruction. To add to the short-term hardships, President Chissano's government, under pressure from the International Monetary Fund, is bent on economic reform and has removed subsidies. The consequential rise in the cost of living for Mozambique citizens has stirred unrest from time to time in Maputo. The government has stressed that the suffering is short-term.

Given permanent political stability, the prospects of eventual well-being are not entirely dismal. The economy is and will continue to be largely agricultural, employing up to 90 per cent of the population, but there is industrial potential in fertilizers, food processing and textiles and there are unexploited reserves of coal, copper, bauxite and offshore gas.

THE REPUBLIC OF MOZAMBIQUE	
STATUS	Republic
AREA	784,755 sq km (302,915 sq miles)
POPULATION	15,583,000
CAPITAL	Maputo
LANGUAGE	Portuguese, tribal languages
RELIGION	majority traditional beliefs, 15% Christian, 15% Muslim
CURRENCY	metical (MZM)
ORGANIZATIONS	Comm., OAU, SADC, UN

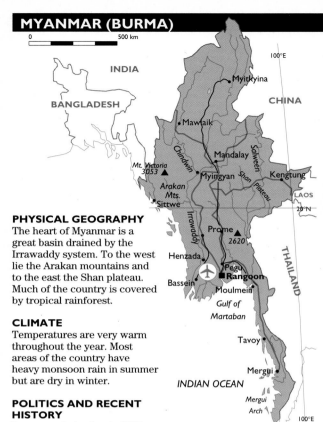

0   500 km

INDIA

BANGLADESH

Mt. Victoria 3053 ▲

Arakan Mts.

CHINA

Myitkyina

Mawlaik

Mandalay

Myingyan

Kengtung

LAOS

Sittwe

Prome 2620

Henzada

Pegu ■Rangoon

Bassein

Moulmein

Gulf of Martaban

Tavoy

Mergui

INDIAN OCEAN

Mergui Arch

THAILAND

100°E

20°N

100°E

PHYSICAL GEOGRAPHY

The heart of Myanmar is a great basin drained by the Irrawaddy system. To the west lie the Arakan mountains and to the east the Shan plateau. Much of the country is covered by tropical rainforest.

CLIMATE

Temperatures are very warm throughout the year. Most areas of the country have heavy monsoon rain in summer but are dry in winter.

POLITICS AND RECENT HISTORY

In a general election in 1990 a coalition National League for Democracy (NLD) secured 80 per cent of the vote. However, the ruling military junta, who call themselves the State Law and Order Restoration Council (SLORC) ignored the result and kept the NLD leader, Aung San Suu Kyi, under house arrest. They had also renamed the country Myanmar in 1989. Work, which started in 1993, on a new constitution has not been completed and it is clear that the ruling junta will insist on a continuing dominant role for the armed forces.

For many years the junta has been opposed by various rebel factions, but it has sought to neutrlise this opposition by either negotiating cease fire agreements as with the Kachin rebels or by inflicting military defeat as it reportedly has

achieved over the Karen National Union amid allegations about the use of chemical weapons. Myanmar's record on human rights remains appalling.

ECONOMY

Myanmar, once rich, is now one of the world's poorest nations, due largely to inept

THE UNION OF MYANMAR	
STATUS	Republic
AREA	678,030 sq km (261,720 sq miles)
POPULATION	444,596,000
CAPITAL	Rangoon (Yangon)
LANGUAGE	Burmese
RELIGION	85% Buddhist. Animist, Muslim, Hindu, Christian minorities
CURRENCY	kyat (MMK)
ORGANIZATIONS	Col. Plan, UN

administration and its isolationist policies. Foreign assistance is suspended but there has been some inward investment from South Korea, Taiwan and Singapore. Moreover Myanmar's representatives have been allowed to attend, a observers, the Association of South Eastern Nations (ASEAN) meetings on the basis that reform is best encouraged by contact rather than isolation. Myanmar's most important trading partner is China who, less concerned than othe nations about human rights, supply arms to the regime, win most of the contracts and import products from Myanmar. There is a potential for wealth based on rice and forestry products although in the latter case illegal logging saps revenue. In addition Myanmar possesses valuable oil and gas reserves and other mineral deposits such as copper.

Most nations are however deterred from investment because of the corrupt regime which in any case expends a high proportion of its resources on the Armed Forces.

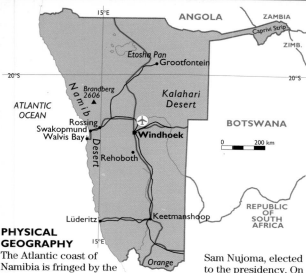

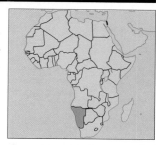

PHYSICAL GEOGRAPHY

The Atlantic coast of Namibia is fringed by the Namib desert. Eastwards, the land rises to a mountainous spine aligned north–south with peaks reaching 2,500 m (8,200 ft). The interior plateau, generally higher than 1,000m (3,300 ft), is the Kalahari desert region.

CLIMATE

Namibia has an arid climate, with any rain falling in summer months. Annual rainfall amounts to less than 50 mm (2 inches) on the coast, rising to 100–250 mm (4–10 inches) in the Kalahari region. Summers are hot and winters cool, with significant diurnal variation because of the high altitude. The coastal strip is kept cool and is subject to fogs because of the effects of the Benguela Current.

POLITICS AND RECENT HISTORY

Namibia, formerly known as Southwest Africa, gained its independence from South Africa in March 1990, following elections to the Assembly in 1989, supervised by the United Nations (UN). The government was formed by the South West African People's Organization (SWAPO) with their leader, Sam Nujoma, elected to the presidency. On gaining independence, Namibia was admitted as a member of the Commonwealth.

Subsequently, Walvis Bay, which had remained a South African enclave, was returned to Namibia and President Nujoma declared a free zone in the port.

The first elections since independence were held in December 1994 and SWAPO gained an overwhelming majority. Their hold on power is such that President Nujoma is able to change the constitution, which he claims favours the opposition, although he has pledged to consult the electorate.

THE REPUBLIC OF NAMIBIA	
STATUS	Republic
AREA	824,295 sq km (318,180 sq miles)
POPULATION	1,461,000
CAPITAL	Windhoek
LANGUAGE	Afrikaans, German, English, regional languages
RELIGION	90% Christian
CURRENCY	Namibian dollar (NAD)
ORGANIZATIONS	Comm., OAU, SADC, UN

ECONOMY

Mineral wealth is the key to the relative prosperity of the Namibian economy. Diamonds provide a significant proportion of the nation's wealth, as does uranium. The Rossin uranium mine is the world's largest. Other minerals extracted include copper, lead and zinc.

Agriculture, sometimes affected by drought, provides revenue, chiefly from beef and the export of pelts from Karakul sheep. The rich coastal waters, fed by the Benguela Current, are the basis of a successful fishing industry. Fish stocks have been depleted because of uncontrolled fishing, especially by Spanish trawlers, but Namibia has imposed strict controls and the stocks are already recovering.

The embryonic manufacturing and tourist sectors are expected to grow and enhance Namibia's prosperity, which is nevertheless highly dependent on trade with South Africa. The distribution of wealth, however, is one of the most imbalanced in Africa. According to a UNICEF (the UN Children's Fund) classification, 55 per cent of the population are 'absolutely poor'.

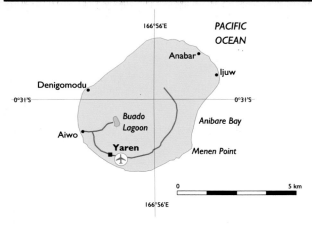

PHYSICAL GEOGRAPHY

Nauru is one of the world's smallest republics, a coral atoll with a fertile strip surrounding a central barren plateau. The highest point reaches only 68m (225 ft) above sea-level. Because of the lack of urban development, Nauru has no official capital. The largest settlement is Yaren.

CLIMATE

The climate is equatorial, modified by sea breezes and the northeast trade winds from March to October. Most of the rainfall occurs between December and March.

POLITICS AND RECENT HISTORY

Independence from United Nations (UN) Trusteeship and Australian administration was granted in 1968. Bernard Dowiyogo was elected president in 1989, defeating Hammer DeRoburt, who was the first head of state and had managed to hold on to power for most of the time since independence.

ECONOMY

In the 1970s Nauru had a higher national income per head than Kuwait, due entirely to the mining of its high-grade phosphates. Today 98 per cent of exports are still phosphates, but reserves are near exhaustion and only a few years of mining remain. The phosphate industry and government service provides the main employment for Nauruans (although 40 per cent of the island's total workforce is composed of overseas contract workers). The island's terrain is damaged beyond repair following 60 years of phosphate extraction, and Nauru is faced with the loss of its only significant asset. Nauru claimed $72 million from Australia to compensate for damage. The claim was settled in 1993, and in 1994 the UK and New Zealand agreed to contribute towards the compensation payments, as the three countries had together set up the British Phosphate Commissioners to run the phosphate industry.

With the impending loss of its vital industry, Nauru has to look elsewhere for income to offset its import bill, since practically all necessities have to be imported, even fresh water. In June 1993 the islanders invested $2 million in staging a London West End musical, *Leonardo*. Unfortunately it met with disastrous reviews and was forced to close after just a month.

Nauru runs an all-Boeing airline service in the Central Pacific region and has also invested in overseas property. Another possibility considered by the government is the creation of an offshore banking system, based on the absence of income tax on the island.

In the last few years, low-lying countries in the Pacific have been the subject of a UN report on the consequences of global warming and the greenhouse effect, threatening a rise in sea-level and flooding. Should this come about, Nauru could almost disappear

THE REPUBLIC OF NAURU	
STATUS	Republic
AREA	21.2 sq km (8 sq miles)
POPULATION	10,000
CAPITAL	Yaren
LANGUAGE	Nauruan, English
RELIGION	Nauruan Protestant majority
CURRENCY	Australian dollar (AUD)
ORGANIZATIONS	Comm. (special member)

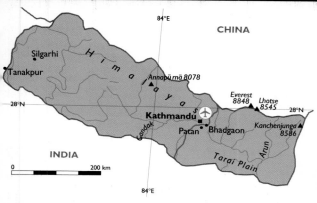

PHYSICAL GEOGRAPHY
The greater part of Nepal is in the Himalayas with access to some of the world's highest mountains, including Everest and Kanchenjunga, both well over 8,500 m (28,000 ft). Southwards, there is a hilly region dissected by valleys. A fertile plain lies along the southern border.

CLIMATE
The climate varies from sub-tropical on the plain to arctic on the peaks. Temperatures in central Nepal vary between 2°C (35°F) and 30°C (86°F). Heavy rainfall, up to 3,000 mm (120 inches) a year, occurs in the mountains in the monsoon season, between January and June

POLITICS AND RECENT HISTORY
From 1962 Nepal was an absolute monarchy, governed by a system of nominated national and local councils, with the king as head of state. However, civil unrest in the early months of 1990 led to a revision of the constitution, with the introduction of a bicameral parliament, and political parties were legalized. In May 1991, the first democratic general elections for 30 years were held and won by the Nepali Congress Party.

In mid-1994, following a year of demonstrations, strikes and allegations of corruption aimed at the ruling party, Prime Minister Girija Koirala was forced to dissolve parliament and call a premature election. In the subsequent poll held in November, the United Communist Party of Nepal won 88 of the 205 parliamentary seats and the Communist leader Man Mohan Adhikari was sworn in as prime minister of a minority government. By June 1995, Adhikari had asked King Birendra to dissolve parliament and call for new elections, in the hope of strengthening the Communist position.

ECONOMY
Nepal is one of the least developed nations in the world, with forestry and subsistence farming occupying 90 per cent of the population. The country is vulnerable to calamity caused by climatic extremes: drought in 1992, followed by torrential rains in 1993, caused floods in

THE KINGDOM OF NEPAL	
STATUS	Kingdom
AREA	141,415 sq km (54, 585 sq miles)
POPULATION	20,812,000
CAPITAL	Kathmandu
LANGUAGE	Nepali, Maithir, Bhojpuri
RELIGION	90% Hindu, 5% Buddhist, 3% Muslim
CURRENCY	Nepalese rupee (NPR)
ORGANIZATIONS	Col. Plan, UN

which bridges were destroyed and 1,600 people died. Despite floods and political disruption, the economy has recently shown growth, with inflation falling from 20 per cent to 7 per cent in 1993, and industrial production, in particular of carpets and garments, soaring with a boom in exports.

Nepal has two major natural assets. The first is its mountains, in particular Everest and Kanchenjunga which, together with the national parks, are attracting increasing tourism. The second is its abundant water resources. However, development projects continue to provoke arguments with environmentalists and participating foreign nations. In 1992–3 the settling of a dispute with India over the Tanakpur project was a dominant issue. Now, a scheme for a further project on the Arun river is awaiting World Bank approval on environmental issues, before being granted support. This project will enable Nepal to sell electric power to India, and will also be a basis for industrial development at home.

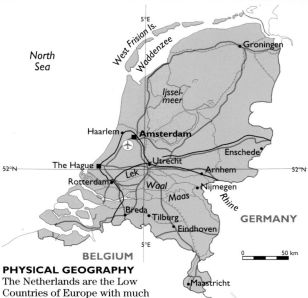

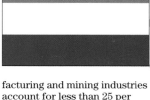

PHYSICAL GEOGRAPHY
The Netherlands are the Low Countries of Europe with much of the land below sea-level and altitudes nowhere exceeding 330 m (1,080 ft). The coastline is often fringed by sand dunes, and inland large tracts of land reclaimed from the sea form the fertile polders.

CLIMATE
The Dutch climate is mild and maritime. Average summer temperatures approach 20°C (68°F) but fall to near 0°C (32°F) in winter, when it is not uncommon for inland waterways to freeze. Rainfall is spread evenly throughout the year.

POLITICS AND RECENT HISTORY
At elections in May 1994, the outgoing prime minister, Ruud Lubbers, stood down in favour of Elco Brinkman. However, Brinkman was unable to command sufficient support in the Dutch parliament and resigned, making way for a coalition of the Labour, right wing Liberal and centre-left D66 parties, under Labour leader Wim Kok.

Dutch politics since World War II have been much focused on Europe and less on the country's former colonial links. The nation is an advocate of European unity, and during its European Union presidency in 1991 was instrumental in convening member states at Maastricht with the aim of further integration in political, economic and social policy.

ECONOMY
Economic prosperity in the Netherlands is heavily based on service industries, which contribute 70 per cent of gross domestic product, while manufacturing and mining industries account for less than 25 per cent. The country is also intensively farmed with both livestock and mixed crops, including cereals, sugar-beet and vegetables.

The Netherlands is a trading nation, with transport equipment and machinery, chemicals and plastics and food, drink and tobacco accounting for most exports.

Although the Dutch economy suffered from the general recession in the early 1990s, particularly that in Germany which is the market for 30 per cent of Dutch exports, the economy is now recovering. The government adopted strict monetary and fiscal measure to protect the guilder's exchange rate within the European exchange rate mechanism.

Plans have been developed to enhance the nation's trading and entrepôt status. A fifth runway is to be built at Amsterdam's Schiphol airport, and approval has been given for a new railway from Rotterdam to the German border, linking the port with the Ruhr.

THE KINGDOM OF THE NETHERLANDS	
STATUS	Kingdom
AREA	41,160 sq km (15,890 sq miles)
POPULATION	15,385,000
CAPITAL	Amsterdam (seat of government – the Hague)
LANGUAGE	Dutch
RELIGION	40% Roman Catholic, 30% Protestant, Jewish minority
CURRENCY	gulden (guilder or florin) (NLG)
ORGANIZATIONS	Council of Europe, EEA, EU, NATO, OECD, UN, WEU

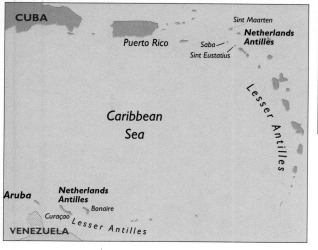

CUBA

Puerto Rico

Sint Maarten

Netherlands Antilles

Saba

Sint Eustatius

Caribbean Sea

Lesser Antilles

Aruba

Netherlands Antilles

Bonaire

Curaçao

Lesser Antilles

VENEZUELA

Netherlands Antilles

Westpunt

69°W

St. Kruis

Ascension

St. Michiel

Curaçao

Emmastad

Willemstad

12° 10'N

New Port

69°W

Caribbean Sea

Bonaire

68°15'W

Rincon

Klein Bonaire

Kralendijk

12° 10'N

ARUBA

STATUS.........**Self-governing Island of Netherlands Realm**

AREA ....**193 sq km**
(75 sq miles)

POPULATION......................**68,897**

CAPITAL......................**Oranjestad**

THE NETHERLANDS ANTILLES

STATUS...........**Self-governing Part of Netherlands Realm**
(includes the islands of Bonaire, Curaço, Saba, Sint Eustatius, Sint Maarten)

AREA ....**993 sq km**
(383 sq miles)

POPULATION **191,311**

CAPITAL......................**Willemstad**

70°W

Arasji

Druif

Oranjestad

Aruba

12°30'N

Sint Nicolaas

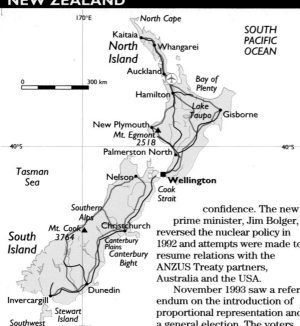

PHYSICAL GEOGRAPHY
On the more heavily populated North Island, mountain ranges, broad fertile valleys and volcanic plateaux predominate. South Island is also mountainous, with the Canterbury Plains forming its only extensive lowland.

CLIMATE
North Island is temperate. South Island has cooler winters and upland snow. Rainfall is distributed throughout the year.

POLITICS AND RECENT HISTORY
New Zealand became fully independent in 1947. Her role in international organizations was affected by the foreign and defence policies of the Labour government, which banned ships powered by nuclear energy or carrying nuclear weapons, from entering the country's waters.

In 1990 Labour was replaced by the National Party, after economic turmoil, two leadership changes and a vote of no confidence. The new prime minister, Jim Bolger, reversed the nuclear policy in 1992 and attempts were made to resume relations with the ANZUS Treaty partners, Australia and the USA.

November 1993 saw a referendum on the introduction of proportional representation and a general election. The voters backed proportional representation in the referendum, while the election resulted in a hung parliament in which Labour and the minor parties did better than expected. However, after the counting of absentee votes from thousands of expatriates, the National Party finally clinched a one-seat majority, returning Bolger to power.

The next general election will be conducted with redrawn constituency boundaries and under the new mixed member proportional system. This system tends to favour small parties, so that some fragmentation of the old order is expected. Meanwhile, the New Zealand government has set aside large sums of money to compensate Maori peoples for loss of land, in a bid to end many years of bitterness surrounding this issue.

ECONOMY
Agricultural products continue to form the basis of the economy, although fish, timber and wood-pulp have become increasingly important commodities, while coal, petroleum and recently natural gas have been successfully exported. Tourism is proving to be New Zealand's fastest-expanding industry, with earnings estimated to reach NZ$9-billion by the year 2000.

The New Zealand economy suffered a serious setback when Britain, its major export market, joined the European Union. Economic reforms became necessary and the country has undergone a long period of restructuring. The welfare systems have been overhauled, subsidies and tariff barriers have been abolished and extensive privatization implemented. The economy is now on a sound footing with steady growth, although inflationary pressures are mounting.

NEW ZEALAND	
STATUS	Commonwealth Nation
AREA	265,150 sq km (102,350 sq miles)
POPULATION	3,493,000
CAPITAL	Wellington
LANGUAGE	English, Maori
RELIGION	35% Anglican Christian, 22% Presbyterian, 16% Roman Catholic
CURRENCY	New Zealand dollar (NZD)
ORGANIZATIONS	ANZUS, Col. Plan, Comm., OECD, UN

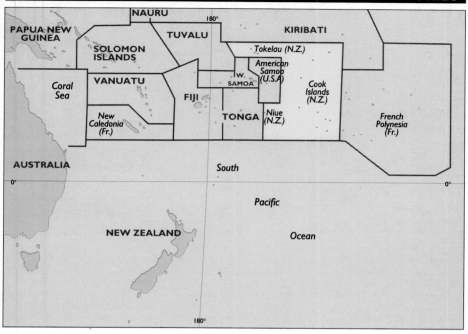

THE COOK ISLANDS

STATUS .... **Self-governing Territory Overseas in Free Association with New Zealand**

AREA .... **233 sq km (90 sq miles)**

POPULATION **19,000**

CAPITAL .... **Avarua**

NIUE

STATUS .... **Self-governing Territory Overseas in Free Association with New Zealand**

AREA .... **259 sq km (100 sq miles)**

POPULATION **2,267**

CAPITAL .... **Alofi**

TOKELAU

STATUS .... **Overseas Territory of New Zealand**

AREA .... **10 sq km (4 sq miles)**

POPULATION **1,690**

CAPITAL .... **each island has its own administrative centre**

NICARAGUA

PHYSICAL GEOGRAPHY
Pine-clad hills separate the east, much of which remains covered with undisturbed forest, from the more developed western regions. The west is dominated by the large Lake Nicaragua. The highest land is in the northwest and the Cordillera Isabella in the north.

CLIMATE
The climate is tropical, with average daily temperatures in excess of 25°C (77°F) throughout the year. On the west coast, wet summer months contrast with a dry period from December to April.

POLITICS AND RECENT HISTORY
Few countries have experienced the extreme political turmoil that has afflicted Nicaragua. The corrupt and dictatorial president Anastasio Somoza, who seized power in 1935, was eventually ousted in 1979, following a lengthy guerrilla war waged by the left wing Sandinistas. The Sandinistas embarked upon a programme of reform but were confronted with armed opposition from the right wing Contras, financed by the US government.

In 1989 a truce was negotiated. In elections the following year, the centrist Violeta Chamorro was victorious with the support of a right wing coalition of 14 parties known as the *Union Nacional Opositora*. President Chamorro suffered defections from the coalition as she sought to bridge the gap between left and right, and was thus without a parliamentary majority. In the early years of her presidency, there were outbreaks of civil trouble, armed conflict and kidnapping of politicians. Happily, following a peace agreement with the majority of the Contra rebels, such events have ceased and the civil war is over. Although President Chamorro had promised to remove Humberto Ortega, brother of the former

THE REPUBLIC OF NICARAGUA	
STATUS	Republic
AREA	148,000 sq km (57,130 sq miles)
POPULATION	4,401,000
CAPITAL	Managua
LANGUAGE	Spanish
RELIGION	Roman Catholic
CURRENCY	gold cordoba (NIC)
ORGANIZATIONS	CACM, OAS, UN

Sandinista president, from his post of Army Commander-in-Chief, he was in fact allowed to stay on until early 1995, to the consternation of the US government in particular. The next presidential elections are due in 1996, and potential candidates are already jockeying for position. The president's son-in-law Antonio Lacayo is a leading contender.

In February 1994 a peace agreement was reached between the government and the last of the rebel Contras.

ECONOMY
The Nicaraguan economy is in chaos. Austere measures required by the International Monetary Fund have brought low inflation but also widespread poverty, more acute than during the Sandinista regime. All agricultural output, the traditional mainstay of the economy, is down, although increases in coffee prices may bring some relief. Around 55 per cent of the population is unemployed. The nation is dependent upon international aid, but US aid worth $98-million was suspended while General Ortega retained his military command. Despite this, the economy is now showing the first signs of improvement.

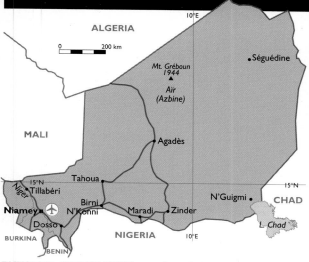

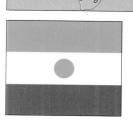

PHYSICAL GEOGRAPHY
Apart from savannah in the south and in the Niger valley, most of the vast country of Niger falls within the Sahara desert. The Aïr mountains in the central region rise from the plains to over 2,000 m (6,562 ft).

CLIMATE
Rainfall is low, with 560 mm (22 inches) in the south, decreasing to virtually zero in the north. Temperatures are high for most of the year, at around 35°C (95°F).

POLITICS AND RECENT HISTORY
Niger, once part of French West Africa, became an independent republic in 1960. A military coup in 1974 introduced rule by a Supreme Military Council, led by a president, and all political groups were banned. In 1991 a five-month national conference ousted the president from his executive position and an interim High Council proclaimed itself the governing body. A new constitution, approved by referendum, was adopted in December 1992 and in the spring of 1993 the first free presidential elections were held. Mahamane Ousmane was elected and in multi-party parliamentary elections the Allied Forces for Change coalition assumed power.

Ousmane's government has been confronted by a series of problems generated by years of drought, recession and interethnic rivalries. The army have threatened mutiny and a prolonged conflict with Tuareg rebels in the north of the country was only brought to an end by a peace deal signed in late 1994. At elections in early 1995, opposition parties gained a one-seat majority over those supporting the president, rendering President Ousmane's tenure precarious.

ECONOMY
Niger's only significant export is uranium, but world prices have continued to slip downwards causing diminishing revenue. Phosphates, coal, tin and tungsten are also mined and other existing minerals are yet to be exploited.

More than 90 per cent of the population is engaged in agriculture, either in livestock rearing or in subsistence farming, with groundnuts and cotton as cash crops. Desertification and drought are ongoing problems, and recently in the north and northeast, an area already short of food, locust swarms contributed to a famine.

Serious pollution of the Niger river has affected many of the 20,000 inhabitants of Tillaberi, causing serious illnesses and some deaths. The cause was traced to a derelict pumping system and poor arrangements for refuse collection. This only served to highlight the need for further foreign aid, on which the country, one of the world's poorest, is already highly dependent.

THE REPUBLIC OF NIGER	
STATUS	Republic
AREA	1,186,410 sq km (457,955 sq miles)
POPULATION	8,361,000
CAPITAL	Niamey
LANGUAGE	French, Hausa, other native languages
RELIGION	85% Muslim, 15% traditional beliefs
CURRENCY	CFA franc (W Africa) (XOF)
ORGANIZATIONS	ECOWAS, OAU, UN

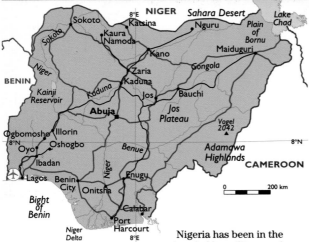

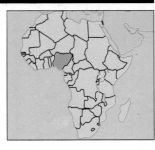

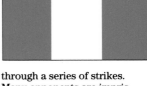

PHYSICAL GEOGRAPHY

A coastal zone of sandy beaches, swamps and lagoons gives way to rainforest, grading into savannah on high plateaus. To the north is the semi-desert edge of the Sahara. The Niger Delta region, together with shallow offshore waters, holds vast underlying reserves of oil and gas.

CLIMATE

The climate is tropical with high temperatures in the region 32°C (90°F) everywhere. The humid south has a heavy rainfall amounting to 2,500 mm (100 inches), falling mainly from April to October. Inland, the rainfall is lower and the rainy season shorter.

POLITICS AND RECENT HISTORY

Nigeria has three major and about 250 smaller ethnic groups. Intense rivalry exists and clashes are always threatening national unity. Especially traumatic was the civil war and famine between 1967 and 1971, when the Ibo-dominated Eastern Region of Nigeria attempted to secede as an independent state of Biafra. Since then, Nigeria has evolved into a federation of 30 states.

Nigeria has been in the main ruled by military regimes since its independence from Britain in 1960. General Ibrahim Babangida assumed the presidency in 1985 but promised to restore civilian rule. After many delays, an election was eventually held in the summer of 1993. The contest was between two candidates and was won by Chief Moshood Abiola. However, Babangida promptly annulled the result, left office and set up an interim administration. In November another military coup installed General Sani Abacha as president.

General Abacha has so far shown no signs of returning to civilian rule and democracy, even though the Nigerian population, especially the Yoruba people of the southwest, have shown their displeasure through a series of strikes. Many opponents are imprisoned without trial. These include Chief Abiola, who was arrested in June 1994 when he declared himself the rightful president. There is much international concern over alleged human rights abuses with Nigeria.

ECONOMY

Virtually all of Nigeria's foreign earnings come from its vast oil resources. Gas reserves are relatively undeveloped, although a pipeline network serving Benin, Togo and Ghana is planned. However, revenue has diminished because of foreign disinvestment, the political turmoil and the threat of sanctions unless there is a return to democracy.

A prestigious natural gas project will not go ahead in the immediate future because of lack of loan funds. A deregulatory budget, declared in early 1995 and evidently aimed at mollifying potential lenders, is unlikely to improve matters.

Corruption is rife within the country. Its oil wealth has been misappropriated, the debt burden is massive, inflation is high and poverty and unemployment have increased.

THE FEDERAL REPUBLIC OF NIGERIA	
STATUS	Federal Republic
AREA	928,850 sq km (356,605 sq miles)
POPULATION	105,264,00
CAPITAL	Abuja
LANGUAGE	English, Hausa, Yoruba, Ibo
RELIGION	Muslim majority, 35% Christian, aminist minority
CURRENCY	naira (NGN)
ORGANIZATIONS	Comm., ECOWAS, OAU, OPEC, UN

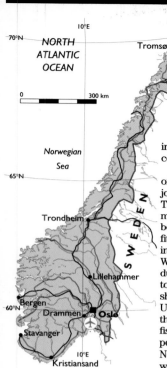

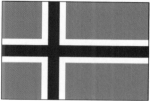

interspersed with periods of coalition rule.

For 25 years, the main issue of debate has been whether to join the European Union (EU). The minority Labour government, led by Gro Brundtland, believe that industry would benefit from competition, more investment and trading partners. When put to the test in a referendum in November 1994, the electorate voted against membership. Among those rejecting the Union were farmers, fearful of the loss of generous subsidies, fishermen anxious to avoid competition from other nations in Norwegian waters and those with interests in the oil industry, who wished to protect home ownership of reserves.

PHYSICAL GEOGRAPHY

The country is characterized by high mountainous terrain and a very long coastline, some 2,000 km (1,250 miles), in the west. To the north are long deep fjords and an offshore island belt, and in the south are forests and many lakes.

CLIMATE

The climate is modified in coastal areas by the North Atlantic Drift. Summers become warmer and winters colder inland. Precipitation, of up to 2,000 mm (80 inches), is distributed throughout the year.

POLITICS AND RECENT HISTORY

Following German occupation during World War II, Norway abandoned a long-standing policy of neutrality and in 1949 joined the North Atlantic Treaty Organization. A minority Labour government has been dominant,

ECONOMY

Twenty years ago, Norway began exploiting the North Sea's vast oil and natural gas resources. Norway is now the biggest oil producer in western Europe, with oil and gas accounting for 40 per cent of its exports. In 1995, production increased to the point where Norway has become the world's second largest net

exporter of crude oil after Saudi Arabia. Further boosts to production will accrue now that the gigantic Troll production platform, the largest and heaviest structure ever moved by man, is in position. Exploitation of natural gas will also increase if plans to build a pipeline to supply France come to fruition.

Two-thirds of the population are engaged in service industries, often related to oil and shipping activities. Tourism, always an important revenue earner, received a boost from the Winter Olympics held in 1994.

Exports include timber, pulp and paper, based on widespread coniferous forests. Hydro-electric power is abundant and cheap, supporting industries such as electrical engineering and aluminium production.

Traditionally, however, the country is a fishing nation, with 90 per cent of its catch exported. Norway is the world's leading producer of salmon, due to fish farming. Whaling has resumed, despite an international ban. It was announced that 232 whales, mainly North Atlantic minke, would be taken in 1995. Norway claim that stocks are adequate to support this quantity, but other countries dispute the figures.

THE KINGDOM OF NORWAY	
STATUS	Kingdom
AREA	323,895 sq km (125,025 sq miles)
POPULATION	4,331,000
CAPITAL	Oslo
LANGUAGE	Norwegian (Bokmal and Nynorsk), Lappish
RELIGION	92% Evangelical Lutheran Christian
CURRENCY	Norwegian krone (NOK)
ORGANIZATIONS	Council of Europe, EEA, EFTA, NATO, OECD, UN

OMAN

PHYSICAL GEOGRAPHY
Oman is a land of deserts, with mountains in Dhofar in the south and in Jebel Akhdar in the north. There are some narrow fertile coastal strips. An enclave in the north of the Musandam peninsula is separated from the remainder of Oman by the territory of the United Arab Emirates.

CLIMATE
The climate is hot and mostly dry, although the coasts are more humid than the interior, and the southern upland regions receive monsoon rains from June to September.

POLITICS AND RECENT HISTORY
Sultan Qaboos bin Said, whose dynasty goes back to the 18th century, is an absolute ruler. He governs by decree and imposes strict censorship. No political parties are permitted. In a gesture towards democracy, a new Consultative Council was established in 1991. Its 59 members are selected by agreement between the palace and regional leaders.

For many years Oman has had an uneasy relationship with its Yemeni neighbours to the south. However, rapprochement in recent years has reduced tension.

ECONOMY
The Omani economy underwent profound change in the late 1960s following exploitation of the nation's oil reserves. Oil rapidly accounted for 99 per cent of export revenue where previously the major commodities had been dates, fish and limes. The oil industry is a state monopoly, although private enterprise is encouraged in other sectors of the economy. Important discoveries of gas have been made and the government has allowed foreign participation in exploiting these reserves.

Mining has been developed in recent years. In 1991 total

THE SULTANATE OF OMAN	
STATUS	Sultanate
AREA	271,950 sq km (104,970 sq miles)
POPULATION	2,018,000
CAPITAL	Muscat (Masqaṭ)
LANGUAGE	Arabic, English
RELIGION	75% Ibadi Muslim, 25% Sunni Muslim
CURRENCY	rial Omani (OMR)
ORGANIZATIONS	Arab League, UN

metal exports reached 11,000 tons. Copper ores from the Sohar region are smelted locally. Similarly, extensive gypsum reserves are utilized in a cement factory in Dhofar.

Agriculture is hampered by water shortages, although a master plan for water supply, involving the construction of dams at wadi mouths, is in development. Meanwhile, over-extraction from aquifers has caused intrusion of sea water and damaged agriculture along the fertile Batinah coast. The main crops are dates, alfalfa, tomatoes, aubergines, limes and, in the south, bananas.

Manufacturing industry makes only a modest 5 per cent contribution to gross domestic product and is confined to traditional output such as silverworking and weaving. There is some scope for tourism, and the government has taken a cautious but positive approach to its development.

The Omani government has been severely criticized by the World Bank for its economic policy – the government's levels of expenditure can not be sustained and unless there is rapid change, there will be the risk of serious social unrest, as wealth from oil and gas declines in the years ahead.

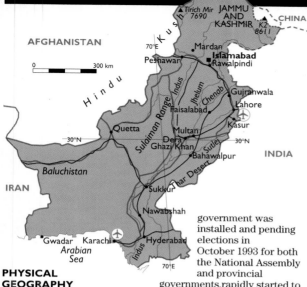

PHYSICAL GEOGRAPHY

Dominated by the great low-land basin drained by the Indus river system, the land rises steeply northwards to the mountains of the Hindu Kush. Baluchistan is semi-desert plateaux and mountain ranges.

CLIMATE

Temperatures are warm, except in the mountains. Rainfall is monsoonal but amounts vary depending on altitude and aspect.

POLITICS AND RECENT HISTORY

In 1971 West and East Pakistan separated, East Pakistan declaring independence as Bangladesh. Pakistani politics have a history of short-term administrations, military influence and corruption. In early 1993 Prime Minister Nawaz Sharif proposed constitutional amendments limiting the power of President Ghulam Ishaq Khan, thus bringing the two men into conflict. In April 1993, the president dismissed the prime minister, but a month later he was reinstated by the Supreme Court. The army intervened and forced the resignation of both men. A caretaker

government was installed and pending elections in October 1993 for both the National Assembly and provincial governments, rapidly started to implement economic and political reforms, the latter aimed at eliminating corruption.

Benazir Bhutto of the Pakistan People's Party won the elections at both national and provincial levels. Together with minor parties, Bhutto was able to form a government and her position was further strengthened when her candidate, Farooq Leghari, won the presidential elections.

ECONOMY

Reforms such as encouragement for overseas investors and privatization of state assets were introduced by the Sharif administration. Further changes, including taxation of land and recovery of debts, were initiated by the caretaker

THE ISLAMIC REPUBLIC OF PAKISTAN

STATUS	Islamic Republic
AREA	803,940 sq km (310,320 sq miles)
POPULATION	122,802,000
CAPITAL	Islamabad
LANGUAGE	Urdu, Punjabi, Sindhi, Pushtu, English
RELIGION	90% Muslim
CURRENCY	Pakistani rupee (PKR)
ORGANIZATIONS	Col. Plan, Comm., UN

government. These reforms have been pursued by Benazir Bhutto, bringing a degree of prosperity, although the budget deficit will call for unpopular spending cuts or revenue increases.

Agriculture remains the most important economic sector, employing about 50 per cent of the workforce and accounting for nearly 70 per cent of export revenue. Wealth, however, is unevenly distributed. Most farmers eke a living from small patches of land while a few wealthy landowners control large acreages. Wheat is grown, but still has to be imported. Rice production is stagnant and the cotton crop, traditionally a valuable export earner, has failed in recent years. More positively, irrigation schemes in northern valleys have produced massively increased yields of crops such as apples and potatoes.

Economic development is dependent on infrastructure improvements: there is a shortage of electrical power and the telecommunications system is poor. Pakistan plans to develop its road system, which will allow potentially lucrative trade with the newly independent Muslim republics of the former Soviet Union.

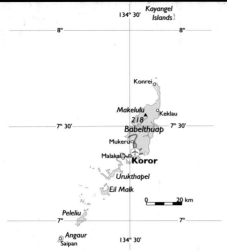

PHYSICAL GEOGRAPHY

The islands, part of the Caroline Islands group, are partly volcanic and partly coral in origin. Within the archipelago of 26 islands and some 300 islets, only 8 islands are inhabited. The total land area is 508 sq km (196 sq mls).

Most people live on the small island of Koror. Koror houses the present capital but new headquarters are being built on the island of Babelthuap.

CLIMATE

The climate is hot and humid throughout the year with heavy rainfall. Temperatures vary little from 32°C - 33°C.

POLITICS AND RECENT HISTORY

Palau, the last of the US-administered UN Trust Territories in the Pacific, became an independent sovereign state on October 1st 1994 and a member of the UN. After nine referenda the island state finally favoured joining the Compact of Free Association with the USA, whereby the US provides defence protection for 50 years and an income over a period of 15 years in return for use of some military facilities.

The eight year delay in finally signing the Compact was due to an anti-nuclear clause in the Palauan constitution which was unacceptable to the US government. The constitution also stipulated that any constitutional amendment could only be made with the support of a 75 per cent majority in a national referendum. This repeatedly failed to be reached and in the event the terms of the referendum were adjusted.

The President, who holds executive power, is Kuniwo Nakamura, while the bicameral National Congress holds legislative authority.

ECONOMY

Fishing and farming are mainly on a subsistence level.

Coconuts, cassava, sweet potatoes, fruit and vegetables are grown. The main cash crop is copra and the traditional chief fish export is tuna. Tourism is an increasingly important source of foreign exchange. Palau is within six hours flying time from the densely populated and prosperous centres of S.E. Asia such as Taiwan and Hong Kong and presently receives about 40,000 visitors a year.

PALAU	
STATUS	Independent in 'free association' with USA
AREA	508 sq km (196 sq miles)
POPULATION	16,000
CAPITAL	Koror
LANGUAGE	English
RELIGION	Christianity (RC)
CURRENCY	US dollar (USD)
ORGANIZATIONS	UN

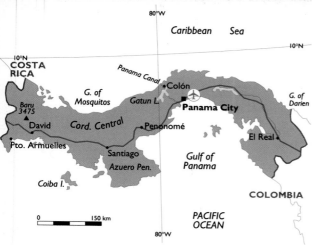

PHYSICAL GEOGRAPHY

Mountain ranges, exceeding 3,000 m (9,800 ft) in height, run the length of the country. Much of the tropical forest has been cleared, but some remains in Darien, towards the border with Colombia.

CLIMATE

Panama has a hot, steamy climate with summer rainfall heaviest along the Pacific coast. The average temperature throughout the year is around 27°C (80°F).

POLITICS AND RECENT HISTORY

Panama seceded from Colombia in 1903, becoming a US protectorate until 1939. The Panama Canal Zone, which was ceded to the USA, came under Panamanian control in 1979 as the Panama Area. The USA continues to exert a powerful political influence on Panama, but will complete the handover of Panama Canal facilities by the end of the century. The Panamanian government has recently established a nine-person management board to oversee the transition of the Canal from the USA.

Following a US invasion in December 1989, the dictator General Noriega was deposed.

President Guillermo Endara Gallimany was installed in his place, after winning the May 1989 elections which had subsequently been annulled. Since then, Noriega has been tried, convicted and imprisoned in the USA for drugs offences, and in October 1993 he was convicted of murder in Panamanian courts.

At the presidential elections of May 1994, Ernesto Balladares won, his party, the Democratic Revolutionary Party, having regained popularity since being driven from power at the time of the American invasion. President Balladares needed only 33 per cent of the vote, because opposition parties were so divided. Shortly after taking office, Balladares, following Costa Rica's lead, abolished the

THE REPUBLIC OF PANAMA	
STATUS	Republic
AREA	78,515 sq km (30,305 sq miles)
POPULATION	2,583,000
CAPITAL	Panama City (Panamá)
LANGUAGE	Spanish, English
RELIGION	Roman Catholic majority
CURRENCY	balboa (PAB), US dollar (USD)
ORGANIZATIONS	CACM, AS, UN

Panamanian army.

ECONOMY

Panama's well-being is founded on revenues from the Canal and from its merchant fleet. The Panamanian fleet is second only to Liberia's and amounts to nearly 50 million tonnes. For many years the economy had been mismanaged and an economic reform programme, entailing privatization and tight monetary policies, was not initiated until Noriega's departure.

Loans from the International Development Bank have been agreed but US-government assistance has not matched Panama's expectations. Even so, the economy expanded following Noriega's downfall, with the construction industry doing particularly well. That expansion may not be sustained – oil exports from Panamanian refineries have declined and banana exports have fallen.

Two infrastructure projects may assist the economy. The Panama Canal is currently being widened to accommodate larger vessels. Plans are also being developed to complete the last section of the Trans-American highway, across the Panama-Colombia border.

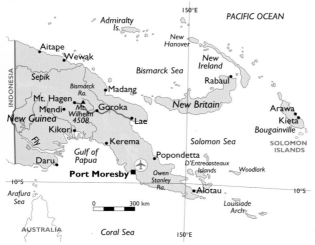

PHYSICAL GEOGRAPHY
In the mountain ranges of central New Guinea, the highest peaks reach well over 4,000 m (13,100 ft). The forested mountains are bordered by plains, swamps and huge delta regions. The country includes some 600 other islands.

CLIMATE
Rain, averaging 2,000 mm (79 inches) annually, falls most heavily between December and March, and there are high temperatures of 21–32°C (70–90°F) all year round.

POLITICS AND RECENT HISTORY
Papua New Guinea was granted full independence from its status as a United Nations Trusteeship under Australian administration in 1975. Government has habitually taken the form of frequently changing coalitions, with the emphasis on individuals rather than political views. The Australian government still maintains a keen interest in the country's affairs.

The 1992 election, in which Paias Wingti won the premiership was, two years later, declared invalid by the Supreme Court. Wingti stepped down and was replaced by the experienced Sir Julius Chan. Within days, Sir Julius had negotiated a cease-fire between General Sam Kauono, the leader of the secessionist Bougainville Revolutionary Army and Papua New Guinea defence forces. Civil war had raged on the island of Bougainville for six years after Bougainvilleans complained of inadequate compensation for land appropriated for mining.

ECONOMY
Papua New Guinea has substantial mineral resources. Until its closure because of civil war, the Panguna copper mine on Bougainville generated nearly half the nation's national income. The mine is now set to resume production. Meanwhile, other mines have been developed. These include Ok Tedi, producing gold and copper, Porgera, producing silver, and most recently Lihir island, said to contain one of the world's largest deposits of gold outside South Africa.

Most of the population is engaged in agriculture, producing cash crops including coconuts, cocoa, coffee, rubber, tea and sugar. However, demand for these products has fallen. In contrast, export of tropical logs has soared in response to rising demand. Papua New Guinea is one of the few countries permitting the export of unprocessed logs. The government, however, has recently announced that this trade is to be phased out by the year 2000. The objective is to ensure the conservation and renewal of forestry resources.

THE INDEPENDENT STATE OF PAPUA NEW GUINEA	
STATUS	Commonwealth Nation
AREA	462,840 sq km (178,655 sq miles)
POPULATION	3,922,000
CAPITAL	Port Moresby
LANGUAGE	English, Pidgin English, native languages
RELIGION	Pantheist majority, Christian minority
CURRENCY	kina (PGK)
ORGANIZATIONS	Col. Plan, Comm., UN

PHYSICAL GEOGRAPHY
The River Paraguay separates two distinct areas: a hilly, heavily forested region to the east and a western zone of marsh and flat alluvial plains.

CLIMATE
The climate is subtropical, with hot, rainy summers from December to March and generally mild winters.

POLITICS AND RECENT HISTORY
For nearly 50 years politics in Paraguay have been heavily influenced by the army in the form of the military-led Colorado Party. For 34 years, from 1954 to 1989, the dictator president, General Alfredo Stroessner, imposed a repressive regime which included states of siege, the expulsion of would-be opposition leaders and disregard for human rights.

In 1989 General Stroessner was deposed in a coup led by General Andrés Rodríguez, who himself became president three months later, promising a civilian successor in 1993. Many of the repressive laws were repealed, although a new constitution, ratified in 1991, still guarantees the autonomy of the army.

In Paraguay's first ever free elections, held in May 1993 amid allegations of electoral malpractice, the civilian candidate of the ruling Colorado Party, Juan Carlos Wasmosy, was victorious. His inauguration as president in August 1993 marked an end to nearly half a century of military rule.

ECONOMY
The Paraguayan economy is based primarily on agriculture, in particular the cash crops of cotton and soya beans, although livestock and forestry products are also important. The lack of a skilled workforce is a major reason for the underdeveloped nature of the economy.

The country is self-sufficient in energy due to its hydro-electric power (HEP) development. The dam at Itaipú, built jointly with Brazil and completed in 1988, is the largest in the world and another massive HEP development is in progress at Yacyretá, this time in conjunction with Argentina. Private capital is being sought for construction of a third, even larger dam at Corpus Cristi.

Paraguay, together with Argentina, Brazil and Uruguay, is a member of Mercosur, the largest common market in Latin America. The four nations have agreed tariffs for trade between themselves and with other countries.

Although the Wasmosy administration has planned economic reform, progress is slow. Inflation is high and unemployment exceeds 10 per cent. The high rate of population growth exacerbates this situation. On the other hand, continued expansion of cash crop output allows Paraguay to record a visible trade surplus.

THE REPUBLIC OF PARAGUAY	
STATUS	Republic
AREA	406,750 sq km (157,005 sq miles)
POPULATION	4,643,000
CAPITAL	Asunción
LANGUAGE	Spanish, Guarani
RELIGION	90% Roman Catholic
CURRENCY	guarani (PYG)
ORGANIZATIONS	Mercosur, OAS, UN

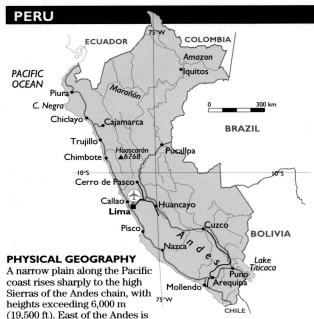

PHYSICAL GEOGRAPHY

A narrow plain along the Pacific coast rises sharply to the high Sierras of the Andes chain, with heights exceeding 6,000 m (19,500 ft). East of the Andes is high plateau country, dissected by fertile valleys, which descends in the extreme east to the Amazon Basin and its tropical rainforests.

CLIMATE

Peru lies in tropical latitudes just south of the Equator but its climate is conditioned by its relief. The coastal plain is cooled by the Humboldt current. The Andes are snow-covered and the interior plateau exhibits wide ranges of temperature. Only in the east of the country do the high rainfall and humidity associated with equatorial latitudes occur.

POLITICS AND RECENT HISTORY

Peru is a presidential republic. President Alberto Fujimori came to power in 1990, following the electoral defeat of Alan Garcia. The presidential term is five years, but President Fujimori's new constitution, permitting re-election, was approved in a referendum in October 1993. Consequently, President Fujimori stood again at presidential elections in April 1995 and easily beat Javez Perez de Cuellar, the former United Nations Secretary General. This victory is no doubt a reward for improved prosperity and the virtual defeat of the left wing Shining Path (*Sendero Luminoso*) revolutionary movement.

The president had, prior to the elections, suspended Congress, ousted opposition parties and announced rule by decree. In order to maintain his authority, he needs the support of the army which exerts considerable influence and has been accused of human rights abuses. Despite this, President Fujimori was able to curb the army in the recent border dispute with Ecuador and adhere to the agreed cease-fire.

THE REPUBLIC OF PERU	
STATUS	Republic
AREA	1,285,215 sq km (496,095 sq miles)
POPULATION	23,088,000
CAPITAL	Lima
LANGUAGE	Spanish, Quechua, Aymara
RELIGION	Roman Catholic majority
CURRENCY	sol (PES)
ORGANIZATIONS	OAS, UN

ECONOMY

During the profligate rule of Alan Garcia the Peruvian economy was in chaos, with inflation reaching nearly 400 per cent per month in 1990. President Fujimori has introduced programmes to deregulate the economy and achieve stabilization. The result is that Peru is the fastest growing economy in Latin America and inflation is significantly reduced, although there is a sizeable budget deficit. Privatization of state-owned enterprises has encouraged foreign investment, especially from Asian countries.

Peru has plentiful natural resources, including copper and gold. The gold mine at Yanacocha is the most productive in Latin America. The cold waters off Peru's coast are rich in marine life and Peru is the world's largest exporter of fish-meal. There are concerns, however, about over-exploitation of this resource.

The illegal growth and export of cocaine is still important to the Peruvian economy, but less so than previously. This is partly due to government action and partly the effect of a fungus that has afflicted the coca bush.

PHYSICAL GEOGRAPHY
Of the numerous islands that make up the Philippines, the larger ones are mountainous and forested. Volcanic activity is common, and the eruption of Mt. Pinatubo in 1991 was one of the most violent this century.

CLIMATE
The climate is tropical, tempered by maritime influence. Rainfall, sometimes accompanied by typhoons, occurs mainly from June to October.

POLITICS AND RECENT HISTORY
Full independence from US commonwealth status was achieved in 1946. When the corrupt dictator Ferdinand Marcos was deposed and fled into exile in 1986, Corazon Aquino, who had contested the election that year, was installed as president. Her political reforms were often stalled by opposition from Congress and the right wing military. Her successor, President Fidel Ramos, although securing only 25 per cent of the vote in the 1992 elections, has been more adept and more popular – proven in congressional elections in May 1995, when the president's supporters gained comfortable majorities in both the Senate and the House of Representatives. The result may be seen as an endorsement for President Ramos' programme of social and economic reform, known as Philippines 2000. It may also be a measure of his success in achieving reconciliation between various factions, such as the army and the Catholic Church. He has not yet quelled Muslim separatists on Mindanao, who periodically carry out terrorist attacks.

ECONOMY
Decades of mismanagement and fraud created a poor nation from one that could have been wealthy. Economic reform was initiated by President Aquino and has gathered pace during the Ramos presidency. Although some overseas aid is still required, the nation is moving towards self-sufficiency and prosperity. This has been achieved by the selling-off of state assets and the encouragement of foreign investment. The result is a buoyant economy and manageable inflation.

Some of the worst infrastructure problems, mainly connected with power supply, are being tackled and the abandonment of the US bases at Subic Bay and Clark Air Force Base, which might have been economically damaging, are being turned to advantage. Subic is being developed as a flourishing free container port while Clark, along with Manila airport, is to be upgraded. The country's shortage of fuel will be partly offset by the discovery of natural gas in the seas west of Palawan Island.

Following his recent electoral victory, President Ramos has announced the next stages of reform, which include reduction of tariffs, simplification of taxes and investment in public health and education.

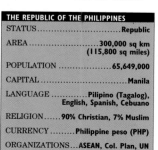

THE REPUBLIC OF THE PHILIPPINES	
STATUS	Republic
AREA	300,000 sq km (115,800 sq miles)
POPULATION	65,649,000
CAPITAL	Manila
LANGUAGE	Pilipino (Tagalog), English, Spanish, Cebuano
RELIGION	90% Christian, 7% Muslim
CURRENCY	Philippine peso (PHP)
ORGANIZATIONS	ASEAN, Col. Plan, UN

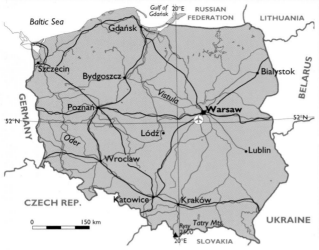

PHYSICAL GEOGRAPHY
Much of Poland lies in the North European plain and is a land of woods and lakes. The highest parts are in the south, where peaks in the Tatry mountains reach almost 2,500 m (8,200 ft).

CLIMATE
The climatic regime is continental, with warm summers and cold winters, when January temperatures fall well below freezing. Most rainfall occurs in the summer months, averaging 520–730 mm (21–29 inches).

POLITICS AND RECENT HISTORY
The Communist regime collapsed in 1989, following pressure from the Solidarity movement based at the Gdansk shipyard, under its leader Lech Walesa. In the 1989 elections, Solidarity's overwhelming success gained them a majority in parliament (the *Sejm*). Lech Walesa became president following electoral success in 1990.

Polish politics have still to gain stability, since the demise of Communism, although a constitutional change in 1993 requiring political parties to win 5 per cent of the popular vote before gaining a seat marked an improvement. In the parliamentary election of September 1993, following this change, Solidarity was eclipsed and a left wing coalition of the Democratic Left Alliance and the Peasant Party formed a government under prime minister Waldemar Pawlak. By early 1995 President Walesa, impatient with lack of progress and possibly with Pawlak's attempts to rebuild relations with Russia, demanded Pawlak's resignation, threatening dissolution of parliament as the alternative. Pawlak resigned and Jozef Oleksy was appointed in his place. These events have rekindled President Walesa's waning popularity such that he may gain a further term of office at the next presidential elections.

ECONOMY
In the years immediately following Solidarity's success in 1989, the Polish economy suffered as the reforms necessary to achieve the transition to a market economy took effect. Prosperity is now beginning to emerge, following several years of economic growth and increasing agricultural output.

Manufacturing output has strengthened and the private sector, in which two million new jobs have been created, is increasing rapidly. On assuming the premiership, Oleksy announced an increase in the rate of the sale of state assets.

On the debit side, unemployment is high – the legacy of overmanning in the Communist era. Inflation, although no longer at the astronomical levels of a few years ago, is still high, although the new zloty was introduced at a rate of one to 10,000 old zlotys.

Despite the tendency to revive links with the East, Poland is now firmly oriented towards western Europe, and has ambitions to join the European Union, having formally applied for membership.

THE POLISH REPUBLIC	
STATUS	Republic
AREA	312,685 sq km (120,695 sq miles)
POPULATION	38,554,000
CAPITAL	Warsaw (Warszawa)
LANGUAGE	Polish
RELIGION	90% Roman Catholic
CURRENCY	zloty (PLZ)
ORGANIZATIONS	Council of Europe, UN

9°W

• Braga

Porto •   Douro

ATLANTIC   Vila Nova
OCEAN   de Gaia
      Aveiro

                ▲ Estrela
                  1991

        Coimbra

)°N

Santarém

      Tagus

.isbon ■
        Barreiro

Setúbal   • Évora

Grândola •

Sines

Lagos

      Faro •   **Gulf of**
9°W   **Cadiz**

0   100 km

SPAIN

Socialist leader, became Portugal's first civilian president for 60 years in 1986 and was re-elected in 1991.

In 1987, Portugal's first majority government was installed. Aníbal Cavaço Silva, the leader of the Social Democratic Party, became prime minister and retained office following electoral success in 1991. At the beginning of 1995, Cavaço Silva stood down from office and was replaced by Fernando Nogueira. Cavaço Silva may now choose to compete in the next presidential elections.

PHYSICAL GEOGRAPHY

North of the River Tagus, most of the country is high ground with pine forests. Land to the south is generally less than 300 m (1,000 ft) above sea-level.

CLIMATE

The north of the country is influenced by the Atlantic and is cool and moist. The south is warmer with dry mild winters.

POLITICS AND RECENT HISTORY

A military coup in 1974 ended nearly four decades of autocratic rule and costly colonial wars. A year later, independence was granted to Portugal's major colonies in Africa. For the next ten years government was dominated by coalitions and minority rule. Dr Mario Soares, the former

ECONOMY

Agriculture, although underdeveloped, employs 12 per cent of the workforce, and wines, cork and fruit are important exports. However, membership of the European Union (EU) has exposed Portuguese agriculture to competition and demonstrated the need for development and investment.

In industry, the chief exports are textiles and clothing, footwear and wood products. The tendency in recent years has been for closure of older industries such as shipbuilding, fertilizers and steel, and the rapid rise of new industry based on foreign investment, such as the industrial complex at Setúbal, south of Lisbon.

Tourism is an important industry. Lisbon was the 1994 European City of Culture and the city will host Expo 98.

Portugal has consistently been the least affluent nation in western Europe and as such receives a massive EU grant for improving its infrastructure. This aid will be doubled between 1994 and 1999.

The Portuguese economy is on an upward curve, despite the effects of recession, and there is a steady increase in per capita gross domestic product. Inflation is under control and unemployment is low by European standards, at 7 per cent .

THE PORTUGUESE REPUBLIC	
STATUS	Republic
AREA	91,630 sq km (35,370 sq miles)
POPULATION	9,868,000
CAPITAL	Lisbon (Lisboa)
LANGUAGE	Portuguese
RELIGION	Roman Catholic majority
CURRENCY	escudo (PTE)
ORGANIZATIONS	Council of Europe, EEA, EU, NATO, OECD, UN, WEU

PORTUGAL – ISLAND REGIONS AND OVERSEAS TERRITORY

MACAU (MACAO)

STATUS .... **Chinese Territory under Portuguese administration (reverts to China 1999)**

AREA .... **16 sq km (6 sq miles)**

POPULATION **374,000**

CAPITAL .... **Macau**

◦ *Corvo*

Santa Cruz • *das Flores*

Flores

28°W

0 ————— 80 kms

Graciosa

A z o r e s

39°N ——————————————————————————————— 39°N

São Jorge • Velas

Terceira

• *Angra do Heroismo*

Faial (Fayal) ▲ Pico *2351* *Pico*

Lajes do Pico

A t l a n t i c   O c e a n

São Miguel

Ponta Delgada ■ *Povoaçã*

AZORES

STATUS ... **Self-governing Island Region of Portugal**

AREA .... **2,335 sq km (901 sq miles)**

POPULATION **237,100**

CAPITAL .... **Ponta Delgada**

28°W

Sta. Maria

MADEIRA

STATUS ... **Self-governing Island Region of Portugal**

AREA .... **796 sq km (307 sq miles)**

POPULATION **253,200**

CAPITAL .... **Funchal**

17°W

Porto Santo

Porto Santo •

33°N ——————————————— 33°N

• *Santana*

P. Ruivo • *1862* ▲

• *Sta. Cruz*

Madeira ■ **Funchal**

Deserta Grande

Bugio

17°W

0 ————— 60 kms

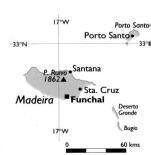

PHYSICAL GEOGRAPHY

The peninsula is flat, never higher than 75 m (250 ft) above sea-level, and barren with sand dunes and salt pans. Vegetation is limited to small patches of scrub.

CLIMATE

The climate is typical of a desert region. Some very limited rainfall occurs in winter, averaging 63 mm (2.5 inches). Summer temperatures are hot and winters are warm. July temperatures average 37°C (98°F).

POLITICS AND RECENT HISTORY

Qatar's ruler, Shaikh Khalifa Bin Hamad Al-Thani, came to the throne in 1972, a year after the state gained its independence from British Protected tatus. He is an absolute monarch and appoints a council of ministers, one of whom is the Crown Prince. The ministers are advised by a 30-member advisory council, but there is no parliament nor any political parties.

The Qatar government is outward looking and in recent years has sought to establish relationships which are not always to the liking of its Gulf

neighbours. It has, for example, planned the construction of a water pipeline from Iran to Qatar, with Iranian co-operation, and has established economic ties with Israel. The state has shown a greater willingness than others in the region to accept Palestinian visitors, despite the Palestine Liberation Organization's (PLO) support for Saddam Hussein during the Gulf War.

Qatar has been plagued by territorial disputes with its neighbours. In September 1992 relations with Saudi Arabia sank to an all-time low following armed clashes over the disputed border. Since then, mediation has achieved a resolution and the relationship has improved. A long-running dispute with Bahrain over the ownership of the reputedly oil-rich Hawar island group has

THE STATE OF QATAR	
STATUS	State
AREA	11,435 sq km (4,415 sq miles)
POPULATION	559,000
CAPITAL	Doha (Ad Dawḥah)
LANGUAGE	Arabic, English
RELIGION	Muslim
CURRENCY	Qatari riyal (QAR)
ORGANIZATIONS	Arab League, OPEC, UN

been referred to the International Court of Justice.

ECONOMY

The Qatari economy has been almost entirely dependent on revenue from hydrocarbons. The lowering of world oil prices has caused sharp fluctuations in the state's gross domestic product, but to some extent the effect has been diminished by the revenue accruing from overseas investments. This revenue is necessary to develop the natural gas reserves in the North Field off Qatar's coast, which holds the world's largest single concentration of natural gas. The initial development phase, completed in 1992, was to supply internal needs, but future exploitation is for export purposes.

A new port is being constructed at Ras Laffan to support the endeavour. Agreement has been reached to supply liquefied natural gas to Japan and deals with Asian countries are under negotiation. Qatar has also agreed, in principle, to supply Israel via a gas pipeline, but has said that the project is conditional on further progress in Middle East peace negotiations.

ROMANIA

PHYSICAL GEOGRAPHY
The dominant physical feature is the Carpathian mountains in the north and centre of the country, which curve around the Transylvanian plateau. To the south lies the valley of the Danube which drains through a delta into the Black Sea.

CLIMATE
Romania enjoys fine warm summers tempered by breezes along the Black Sea coast. Winters are cold, especially in the mountains.

POLITICS AND RECENT HISTORY
In 1989 Romania emerged from a bleak period of corrupt Communist dictatorship when Nicolae Ceausescu, who had ruled for 24 years, was deposed and executed. The revolution had been led by the National Salvation Front, and their leader Ion Iliescu became interim president. Since then, the National Salvation Front, now renamed the Party of Social Democracy, has remained the largest single party with Ion Iliescu as president and Nicolae Vacaroiu as prime minister. However, the party controls only one-third of the parliamentary seats and has com-

bined with several nationalist groups in order to govern. Through this coalition, the party has been able to stave off a series of no confidence motions from the more reformist-minded opposition.

Romania has a large Hungarian minority population, whose treatment has been a source of tension with Hungary. Some concessions such as Hungarian language teaching and Hungarian street signs have been made.

ECONOMY
The Romanian economy has not made a successful emergence from Communism. Its starting position was worse than some other countries: Ceausescu's rule was more centralized and indulged in more foolhardy projects than

ROMANIA	
STATUS	Republic
AREA	237,500 sq km (91,699 sq miles)
POPULATION	22,742,000
CAPITAL	Bucharest (Bucureşti)
LANGUAGE	Romanian, Magyar
RELIGION	85% Romanian Orthodox
CURRENCY	leu (ROL)
ORGANIZATIONS	Council of Europe, UN

elsewhere. In addition the ruling party, composed mainly of previously Communist officials has shown less enthusiasm for economic reform than other regimes.

However, there are signs of progress, prompted in part by the International Monetary Fund which resumed loans to Romania in 1994. Progress has, however, been hampered by the coal miners who have, on several occasions, sought to use their industrial power to preserve the status quo.

The government has succeeded in reducing inflation and has started to attract foreign investment. The sale of 50 per cent of state enterprises, via a voucher system, was announced in mid-1995.

An economic curiosity recently swept Romania in its search for prosperity. Millions subscribed to a pyramid selling organization known as Caritas. Despite a few making a profit, most, inevitably, lost their investment and in August 1994 the organiser of Caritas was arrested.

PHYSICAL GEOGRAPHY

The great size of the Russian Federation means that within its borders there is enormous physical diversity. The highest mountain ranges, such as the Caucasus and Altai, are found along its southern borders but much of the remainder is plain, steppe and plateaux, often forest covered. The great river systems of Siberia, such as the Ob, Yenisei and Lena drain into the Arctic Ocean. West of the Urals the major river, the Volga, runs southwards to the Caspian Sea.

CLIMATE

In general, Russia lies within temperate latitudes, although its more northerly regions are sub-arctic. The climate throughout most of the territory shows continental characteristics, with large seasonal variations in temperature and precipitation occurring mainly in the summer months.

POLITICS AND RECENT HISTORY

Russia, as it is presently constituted, emerged from the collapse of the Soviet Union at the end of 1991. Boris Yeltsin,

elected President of Russia in June 1991, who had defended the Russian parliament against the abortive coup in August of that year, was the legitimate leader of the new nation which took the place of the Soviet Union at the United Nations. In early 1992, Yeltsin was instrumental in establishing the Commonwealth of Independent States, a loose economic grouping which most of the former republics agreed to join.

Throughout 1992 and the greater part of 1993 the dominant political issue within Russia was the conflict between Yeltsin and his reformist advisors.This culminated in the infamous siege of parliament, the arrest of

Alexander Rutskoi and Ruslan Khasbulatov and the survival of Yeltsin's presidency.

Yeltsin had promised new elections, which were duly held in December 1993. As well as electing members of parliament, Russians also voted on a new constitution which gave increased powers to the president, including the right to form and dismiss government and parliament.

Yeltsin won approval for the constitution but he did not get the parliment he sought. Reformist deputies are now outnumbered by anti-reformist elements. The most notorious of these is the fascist and incongruously named Liberal Democratic Party, under the leadership of Vladimir Zhirinovsky. Since the elections, many of the key personalities who promoted reform have either resigned or been eased out of office.

It is hardly surprising that within Russia, which embraces a wide diversity of peoples, ethnic tensions should arise. The most recent trouble has been in the Caucasus republic of Chechnia. Under their leader

THE RUSSIAN FEDERATION	
STATUS	Federation
AREA	17,078,005 sq km (6,592,110 sq miles)
POPULATION	148,366,000
CAPITAL	Moscow (Moskva)
LANGUAGE	Russian
RELIGION	Russian Orthodox, Jewish and Muslim minorities
CURRENCY	rouble
ORGANIZATIONS	Council of Europe, CIS, UN

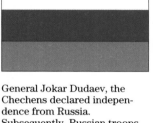

General Jokar Dudaev, the Chechens declared independence from Russia. Subsequently, Russian troops invaded in an effort to suppress the rebellion. This did not succeed and fighting continues. Even though these actions have attracted widespread disapproval, President Yeltsin appears set on defeating the Chechens by military rather than diplomatic means. Yeltsin will need to find a solution if he is to retain the presidency in the next elections, when Zhirinovsky is likely to oppose him.

ECONOMY

The challenge for Russia is to achieve the transition from the state-planned, state-run, centralist economy of the Communist era to one which is market orientated. So far progress has been patchy, due largely to political turmoil. Many Russians live in poverty, and real wages are much reduced. At times, it seemed that economic reforms were beginning to work. Tight fiscal policies had reduced monthly inflation to single figures in August 1992, but the giant,

inefficient industries of the Communist era were suffering. Rather than let them die, the government came to the rescue by increasing their money supply, once more fuelling inflation.

Worse still, inept monetary control led to a drastic collapse of the rouble in September and October 1994, although some of its value was restored shortly thereafter.

Nevertheless, there have been some encouraging reforms. Approximately 60 per cent of Russian industry is now in private hands and the International Monetary Fund has been sufficiently impressed to agree a $6.4-billion loan.

Russia needs to attract more foreign investment, especially into its oil industry. Reserves are massive, but exploitation is so poor that the country suffers from petrol shortages.

Overall, economic change needs to gather pace. If not, the large gap between wealthy and poor will start to create severe social tensions.

REPUBLIC OF ADYGEA
STATUS	Republic
AREA	7,600 sq km (2,934 sq miles)
POPULATION	437,000
CAPITAL	Maykop

REPUBLIC OF ALTAY
STATUS	Republic
AREA	92,600 sq km (35,740 sq miles)
POPULATION	196,000
CAPITAL	Gorno-Altaysk

REPUBLIC OF BASHKORTOSTAN
STATUS	Republic
AREA	143,600 sq km (55,430 sq miles)
POPULATION	3,984,000
CAPITAL	Ufa

REPUBLIC OF BURYATIA
STATUS	Republic
AREA	351,300 sq km (135,650 sq miles)
POPULATION	1,056,000
CAPITAL	Ulan-Ude

STATUS	Republic
AREA	10,300 sq km (5,000 sq miles)
POPULATION	700,000
CAPITAL	Grozny

CHUVASH REPUBLIC
STATUS	Republic
AREA	18,300 sq km (7,064 sq miles)
POPULATION	1,346,000
CAPITAL	Cheboksary

REPUBLIC OF DAGESTAN
STATUS	Republic
AREA	50,300 sq km (19,416 sq miles)
POPULATION	1,854,000
CAPITAL	Makhachkala

REPUBLIC OF INGUSHETIA
STATUS	Republic
AREA	9,000 sq km (3,500 sq miles)
POPULATION	500,000
CAPITAL	Nazran

KABARDINO-BALKAR REPUBLIC

STATUS .... **Republic**

AREA .... **12,500 sq km**
(4,825 sq miles)

POPULATION **777,000**

CAPITAL .... **Nal'chik**

REPUBLIC OF KALMYKIA-KHAL'MG TANGCH

STATUS .... **Republic**

AREA .... **75,900 sq km**
(29,300 sq miles)

POPULATION **328,000**

CAPITAL .... **Elista**

KARACHAY-CHERKESS REPUBLIC

STATUS .... **Republic**

AREA .... **14,100 sq km**
(5,442 sq miles)

POPULATION **427,000**

CAPITAL .... **Cherkessk**

REPUBLIC OF KARELIA

STATUS .... **Republic**

AREA .... **791,000 sq km**
(172,400 sq miles)

POPULATION **791,000**

CAPITAL .... **Petrozavodsk**

REPUBLIC OF KHAKASSIA

STATUS .... **Republic**

AREA .... **61,900 sq km**
(23,855 sq miles)

POPULATION **577,000**

CAPITAL .... **Abakan**

KOMI REPUBLIC

STATUS .... **Republic**

AREA .... **415,900 sq km**
(160,540 sq miles)

POPULATION **1,265,000**

CAPITAL .... **Syktyvkar**

REPUBLIC OF MARI EL

STATUS .... **Republic**

AREA .... **23,200 sq km**
(8,955 sq miles)

POPULATION **758,000**

CAPITAL .... **Yoshkar-Ola**

MORDOVIAN REPUBLIC

STATUS .... **Republic**

AREA .... **26,200 sq km**
(10,110) sq miles)

POPULATION **964,000**

CAPITAL .... **Saransk**

NORTH OSSETIAN REPUBLIC

STATUS .... **Republic**

AREA .... **8,000 sq km**
(3,088 sq miles)

POPULATION **643,000**

CAPITAL .... **Vladikavkaz**

REPUBLIC OF SAKHA (YAKUTIA)

STATUS .... **Republic**

AREA .... **3,103,200 sq km**
(1,197,760 sq miles)

POPULATION **1,109,000**

CAPITAL .... **Yakutsk**

REPUBLIC OF TATARSTAN

STATUS .... **Republic**

AREA .... **68,000 sq km**
(26,250 sq miles)

POPULATION **3,679,000**

CAPITAL .... **Kazan**

REPUBLIC OF TUVA

STATUS .... **Republic**

AREA .... **170,500 sq km**
(65,810 sq miles)

POPULATION **307,000**

CAPITAL .... **Kyzyl**

UDMURT REPUBLIC

STATUS .... **Republic**

AREA .... **42,100 sq km**
(16,250 sq miles)

POPULATION **1,628,000**

CAPITAL .... **Izhevsk**

RWANDA

PHYSICAL GEOGRAPHY
A ridge of volcanic peaks, 3,000 m (9,800 ft) in height, traverses the country from north to south.

CLIMATE
The climate is warm, with a dry season from June to August.

POLITICS AND RECENT HISTORY
Formerly part of Ruanda-Urundi, Rwanda gained independence in 1962. Civil war has raged since 1990, when the Rwanda Patriotic Front (RPF) invaded from Uganda. The origin of the conflict is tribal: the RPF represents the minority Tutsi people, ousted from the government in the 1960s.

In April 1994 the presidents of Rwanda and Burundi were killed, when their plane was shot down as they returned from a meeting to discuss ending the ethnic troubles. The majority Hutu government was ousted by the RPF. Intense fighting between the two rival tribes followed, causing thousands of deaths and creating more refugees. Further massacres were reported in early 1995. The small United Nations peace-keeping contingent has been able to do little to stop the genocide.

ECONOMY
Rwanda is densely populated, surviving by intense cultivation. However, the civil war has brought such turmoil that some three million Rwandese face starvation and a massive injection of aid is needed. Coffee, once its only significant export crop, no longer brings in any worthwhile revenue.

THE REPUBLIC OF RWANDA	
STATUS	Republic
AREA	26,330 sq km (10,165 sq miles)
POPULATION	7,554,000
CAPITAL	Kigali
LANGUAGE	French, Kinyarwanda (Bantu), tribal languages
RELIGION	50% animist, Christian (mostly Roman Catholic)
CURRENCY	Rwanda franc (RWF)
ORGANIZATIONS	OAU, UN

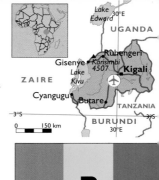

ST KITTS AND NEVIS

PHYSICAL GEOGRAPHY
The pair of volcanic islands each have peaks exceeding 1,000 m (3,300 ft). Their slopes are covered with tropical forest.

CLIMATE
The climate is warm, but tempered with trade winds. Rainfall is abundant throughout the year, the driest period being spring.

POLITICS AND RECENT HISTORY
St Kitts and Nevis gained full independence in 1983. The territory had included the island of Anguilla but, following an abortive declaration of independence in 1967, that island was separated and remains a British dependency. Following elections in 1993, the People's Action Movement, led by Kennedy Simmonds, formed a coalition government with the Nevis Reformation Party.

ECONOMY
The economic mainstay of St Kitts is sugar, which accounts for 60 per cent of export value, and most arable land is devoted to cane plantations. Because of declining world prices in this commodity, injections of aid have been necessary to maintain stability. Other crops contributing to the economy are cotton, especially on Nevis, and a variety of tropical fruits.

In recent years, tourism has developed and there has been growth in light manufacturing and service industries, for example the production of electronic components and data processing.

THE FEDERATION OF ST KITTS AND NEVIS	
STATUS	Commonwealth State
AREA	261 sq km (101 sq miles)
POPULATION	42,000
CAPITAL	Basseterre
LANGUAGE	English
RELIGION	Christian (mainly Protestant)
CURRENCY	E Caribbean dollar (XCD)
ORGANIZATIONS	Caricom, Comm., OAS, UN

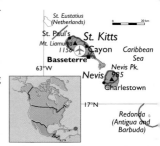

PHYSICAL GEOGRAPHY
St Lucia is volcanic and features forested mountains, the highest of which approaches 1,000 m (3,300 ft).

CLIMATE
Temperatures are hot and tempered by trade winds. Rainfall, always abundant, is heaviest in summer and autumn.

POLITICS AND RECENT HISTORY
St Lucia was granted indepen-

ST LUCIA	
STATUS	Commonwealth State
AREA	616 sq km (238 sq miles)
POPULATION	139,000
CAPITAL	Castries
LANGUAGE	English, French patois
RELIGION	82% Roman Catholic
CURRENCY	E Caribbean dollar (XCD)
ORGANIZATIONS	Caricom, Comm., OAS, UN

dence from Britain in 1979. Democratic elections are held every five years. In 1992 the United Workers Party won with an overall majority of five seats. Their leader, John Compton, was appointed prime minister, winning a third consecutive five-year term of office.

ECONOMY
Agriculture dominates the economy, with light industry and tourism also important. Commercial developments may, in future, add to the island's wealth. A \$75-million US aid programme supported the construction of a container port, opened in 1993, and new facilities have been established at the international airport. The government plans a free zone to encourage trans-shipment trade.

St Lucia is the largest producer of bananas in the

Windward Islands and was, therefore, the most severely afflicted when the crop was devastated by a tropical storm in September 1994. Production has been further set back by farmers, who want a restructuring of the industry.

PHYSICAL GEOGRAPHY
Of volcanic origin, St Vincent itself is forested and boasts one of the two active volcanoes in the eastern Caribbean, Soufrière.

CLIMATE
St Vincent has a maritime tropical climate with warm temperatures throughout the year. The heaviest rainfalls are in summer and autumn.

POLITICS AND RECENT HISTORY
St Vincent gained indepen-

ST VINCENT AND THE GRENADINES	
STATUS	Commonwealth State
AREA	389 sq km (150 sq miles)
POPULATION	110,000
CAPITAL	Kingstown
LANGUAGE	English
RELIGION	Christian
CURRENCY	E Caribbean dollar (XCD)
ORGANIZATIONS	Caricom, Comm., OAS, UN

dence in 1979. In the 1989 elections the New Democratic Party (NDP) retained office when it won all 15 seats in the House of Assembly. A general election in early 1994 returned the NDP, led by James Mitchell, to office for a third term. In September 1994, the two opposition parties merged to form the United Labour Party, under leader Vincent Beache.

ECONOMY
The economy is dependent on agriculture and tourism, which together contribute almost 50 per cent of gross domestic product. The principal crop is bananas, mostly exported to the UK. However, because of overdependence on this commodity, diversification has recently been promoted. Arrowroot, used in medicines and the manufacture of computer paper, is also important.

Tourism is well established on both the main island and the

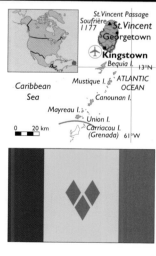

Grenadines. On Union Island, in the south, a \$100-million US-financed tourist development project is planned. It will include a marina and luxury hotel and will no doubt boost the economy.

SAN MARINO

PHYSICAL GEOGRAPHY
The hilly landscape culminates in the three peaks of Mt. Titano, at 739 m (2,425 ft).

CLIMATE
The climate is temperate, with some snow in winter and occasional showers during summer.

POLITICS AND RECENT HISTORY
San Marino is the only surviving Italian city-state, and the world's smallest republic.

THE REPUBLIC OF SAN MARINO	
STATUS	Republic
AREA	61 sq km (24 sq miles)
POPULATION	24,000
CAPITAL	San Marino
LANGUAGE	Italian
RELIGION	Roman Catholic
CURRENCY	Italian lira (ITL), San Marino coinage
ORGANIZATIONS	Council of Europe, UN

Since 1862, various treaties of friendship and co-operation have been signed with Italy. San Marino has no armed forces, and the police are hired from Italy. The state has its own legal system. Legislative power is vested in an elected council of 60 members residing over nine 'castles' – the nine original parishes. A left-wing coalition governs, dominated by the Christian Democrats.

ECONOMY
Tourism is the main source of foreign exchange; there are over 2.5 million visitors a year. Agriculture is based on livestock, wheat and grapes. The main export is wine, while other exports include building stone, ceramics, textiles and furniture. Sales of postage stamps contribute 10 per cent of the national income.

San Marino has close economic ties with Italy. In return

for rules on foreign exchange and for no customs restrictions at its borders, San Marino receives annual subsidies from the Italian government.

SÃO TOMÉ AND PRÍNCIPE

PHYSICAL GEOGRAPHY
The two main islands are mountainous and tree covered.

CLIMATE
Temperatures are consistent with maxima approaching 30°C (86°F) throughout the year. Rainfall is plentiful, but July and August are dry.

POLITICS AND RECENT HISTORY
Following independence from Portugal in 1975, the São Toméan government adhered to Marxist principles. In 1990 a

SÃO TOMÉ AND PRÍNCIPE	
STATUS	Republic
AREA	964 sq km (372 sq miles)
POPULATION	122,000
CAPITAL	São Tomé
LANGUAGE	Portuguese, Fang
RELIGION	Roman Catholic majority
CURRENCY	dobra (STD)
ORGANIZATIONS	OAU, UN

new constitution, allowing for opposition parties, was introduced and in free elections in 1991 the Party for Democratic Convergence (PCD) won. The opposition leader, Miguel Trovoada, was later elected unopposed as president. In legislative elections in October 1994, the PCD were defeated by the MLSTP-PSD, who had been defeated in the first elections in 1991.

ECONOMY
The economy is heavily dependent on cocoa, which provides 90 per cent of revenue. There is, however, a high dependence on imports, one-third of which are food products.

The second most important revenue earning item is the issuing of fishing permits, mainly to European countries. Privatization is being introduced but some diversification is necessary, perhaps through tourism. There is also encour-

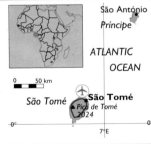

agement for greater investment in coffee and copra, the latter being the main product in the south where yields have jumped in recent years. Industry is minimal, confined to factories producing beer, soap, textiles and processed food.

PHYSICAL GEOGRAPHY
Most of Saudi Arabia's territory is desert or semi-desert plateaux. Towards the west ranges of mountains reaching heights well over 3,000 m (9,800 ft) run parallel to the Red Sea and drop in a steep scarp to a coastal plain known as the Tihama.

CLIMATE
The temperature in summer is hot, with figures above 40°C (104°F) common; winters are warm. Rainfall is at best slight and, in the desolate Empty Quarter, virtually non-existent.

POLITICS AND RECENT HISTORY
Saudi Arabia emerged as a sovereign kingdom in the years between the two world wars, when the Saud dynasty extended its territory. The form of government is an absolute monarchy. King Fahd ibn Abdul Aziz al Saud succeeded in 1982 upon the death of his brother. As well as monarch, he is also prime minister and appoints a council of ministers.

In August 1993 King Fahd met a longstanding commitment to set up an advisory council. This body, known as the *Majlis-al-Shura*, is composed of 60 individuals selected by the king, and its role is to provide advice to the king and his ministers on issues referred to it. There is, however, a growing clamour within Saudi Arabia, both from young intellectuals and Muslim fundamentalists, for a greater degree of democracy.

The Gulf War left its scars on the Saudi political scene and caused a cooling in the relationship between the Gulf States and the Palestine Liberation Organization, who supported Iraq. However, an agreement has recently been signed between Saudi Arabia and Yemen, resolving a dispute over their common border. Saudi Arabia has been a strong supporter of the Middle East peace process and has offered substantial sums to assist Palestinians in the West Bank territories.

ECONOMY
The popular image of Saudi Arabia as an extremely wealthy nation is not entirely true. It does have oil wealth, amounting to a quarter of the world's known reserves, and it can expect to extract at least eight million barrels a day for the next century and probably beyond. Despite this, a budget deficit which has existed for 10 years is growing, and it is unlikely that oil prices will rise to rectify matters. This situation has prompted severe warnings from the International Monetary Fund and in response the Saudi government has imposed budgetary cuts.

This alone is unlikely to be sufficient. It will probably be necessary for the government to withdraw some of the generous subsidies, and perhaps even raise taxes and allow some state assets to pass into private hands. Until now, such measures have been regarded as socially and politically unacceptable.

THE KINGDOM OF SAUDI ARABIA	
STATUS	Kingdom
AREA	2,400,900 sq km (926,745 sq miles)
POPULATION	17,119,000
CAPITAL	Riyadh (Ar Riyād)
LANGUAGE	Arabic
RELIGION	90% Sunni Muslim, 5% Roman Catholic
CURRENCY	Saudi riyal (SAR)
ORGANIZATIONS	Arab League, OPEC, UN

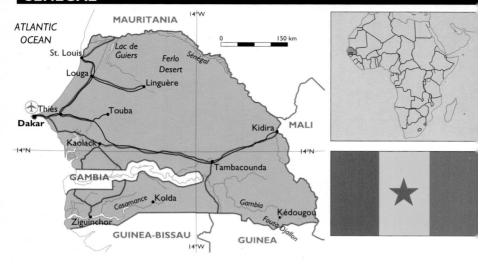

PHYSICAL GEOGRAPHY
Arid semi-desert covers the north of Senegal, while the south is mainly fertile savannah bushland. In the southeast, plains rise up to the Fouta Djallon foothills.

CLIMATE
A tropical climate prevails, with humid rainy conditions between June and October and a drier season falling between December and May. The annual rainfall averages 560 mm (20 inches) and the temperatures range from 22–28°C (72–82°F).

POLITICS AND RECENT HISTORY
After years of French influence and a brief federation with Mali (1959–60), Senegal became a separate independent republic in 1960. Despite the introduction of a multi-party system in 1974, the left wing *Parti Socialiste* remains in power after 33 years. In the elections of May 1993, however, the party returned with a reduced majority. In March 1995, at the request of President Abdou Diouf, Prime Minister Habib Thiam announced the formation of a coalition government with the main opposition, the

Senegalese Democratic Party, in the hope of settling political instability.

Over the past decade Senegal has been subject to fluctuating relations with neighbouring countries. In 1982 the Senegambian Confederation was established with The Gambia, but disputes over trade led to its break-up in 1989. Problems with Mauritania followed – violent ethnic clashes occurred, but diplomatic relations were eventually restored in 1992. Relations with Guinea-Bissau remain strained after territorial quarrels over land which may yield oil.

Government and separatist rebels have clashed in the southern Casamance province. Although a cease-fire was signed

in July, the fighting continues.

ECONOMY
Over 70 per cent of the population is involved in agriculture. Groundnuts, the principal export crop, account for 40 per cent of all cultivated land. However, the land is prone to desertification and a huge yield variation is attributed to the pattern and amount of rainfall.

Diversification into fish, phosphates, textiles and petroleum products is under way, and Senegal has more industry than many west African states. However, protectionist trade policies have made Senegal uncompetitive in world markets. Unemployment has become a problem, with a rate of 24 per cent estimated in Dakar in recent years.

Senegal was regarded by some as being on the verge of bankruptcy, weighed down by internal debts and foreign debt payment arrears. In March 1994 a loan from the International Monetary Fund (IMF) was agreed. The value of the currency was cut by 50 per cent in January 1994 and the IMF called for external aid.

THE REPUBLIC OF SENEGAL	
STATUS	Republic
AREA	196,720 sq km (75,935 sq miles)
POPULATION	8,152,000
CAPITAL	Dakar
LANGUAGE	French, native languages
RELIGION	94% Sunni Muslim, animist minority
CURRENCY	CFA franc (W Africa) (XOF)
ORGANIZATIONS	ECOWAS, OAU, UN

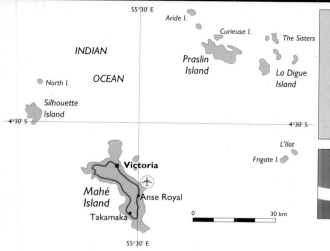

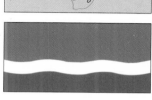

PHYSICAL GEOGRAPHY

The Seychelles is an archipelago of 115 islands in the Indian Ocean, the majority of which are mountainous granite. The remainder are low-lying uninhabited coral atolls. Eighty-eight per cent of the population lives on Mahé.

CLIMATE

The climate is hot and humid all year round, with temperatures in the region of 26–27°C (78–80°F). Rain, averaging 2,400 mm (95 inches) a year, falls mainly between December and May.

POLITICS AND RECENT HISTORY

For most of the time since independence from Britain in 1976, the Seychelles has been ruled by a one-party system under the leadership of Albert René. In 1991 political parties were legalized, and there are now seven. In a multi-party poll, held in July 1992 to elect a 20-person commission to draft a new constitution, President René's ruling Seychelles People's Progressive Front (SPPF) took 14 seats. The first draft constitution failed to reach approval at a referendum in November 1992, and in January 1993 the commission reconvened. The first multi-party elections for 16 years were held in July 1993. President René and the SPFP party defeated the main opposition led by Sir James Mancham, the original president of Seychelles who was ousted by a coup in 1977.

ECONOMY

An economic boom in recent years can be attributed to the success of tourism, based on the appeal of white sands and exotic flora and wildlife.

Farming is restricted on the islands because of poor soil and uneven terrain and there is a heavy dependence on imported food – all domestic requirements have to be imported.

Traditional products of the Seychelles such as copra and cinnamon, as well as the other cash crops of vanilla and tobacco, have declined over the last two decades, not only because of the relatively high costs of production, but also because of competition from fruit and vegetables for the limited amount of usable land. The tea industry has shown signs of revival. Tuna fishing, although mainly artisanal, provides employment for many, and canned tuna has become a useful export.

The Seychelles has a considerable trade deficit, since exports cover less than one-third of the import bill. Nevertheless, the economy is growing on the strength of tourism, and unemployment is nil.

At one time the nation's economy was linked to African countries such as Tanzania. In recent years, however, the trend has been an orientation towards the rich Gulf states, such as Dubai and Singapore.

THE REPUBLIC OF SEYCHELLES	
STATUS	Republic
AREA	404 sq km (156 sq miles)
POPULATION	72,000
CAPITAL	Victoria
LANGUAGE	English, French, Creole
RELIGION	92% Roman Catholic
CURRENCY	Seychelles rupee (SCR)
ORGANIZATIONS	Comm., OAU, UN

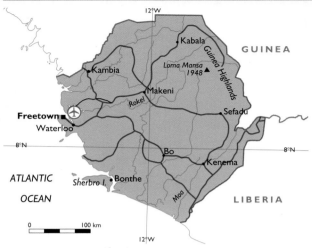

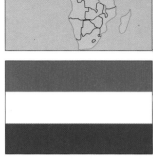

PHYSICAL GEOGRAPHY
A flat plain 113 km (70 miles) wide stretches the length of the coast. Beyond it there is a central forested area rising to the Guinea Highlands.

CLIMATE
The climate is tropical, hot throughout the year with a pronounced humid, rainy season between May and November. Coastal areas are cooled by sea breezes.

POLITICS AND RECENT HISTORY
Sierra Leone became independent from Britain in 1961, and declared itself a republic in 1971. For more than two decades the country had been controlled by the All Peoples' Congress (APC), and from 1978 this was the sole legal party. In a 1991 referendum, 60 per cent of the electorate voted in favour of introducing multi-party democracy, but in April 1992 the APC government was ousted in a military coup led by Captain Valentine Strasser, who set up a governing National Provisional Ruling Council (NRPC).

Civil war has now erupted in Sierra Leone. The Revolutionary United Front, under its leader Foday Sankoh, is waging an armed struggle against government forces. Captain Strasser has offered a cease-fire and unconditional talks and has encouraged the dissidents to participate in elections. They have, however, refused, demanding that the military rulers resign and hand power to a national conference, who will decide the nation's future. Meanwhile, some 900,000 people are displaced.

ECONOMY
Seventy-five per cent of the population is employed in subsistence farming. The main cash crops are cocoa, coffee, ginger, nuts and palm kernels. Mining, however, is the most important source of foreign exchange for the country. Diamonds, gold, bauxite and iron ore are mined and exploitation of diamonds is now to be extended to the sea bed. Sierra Leone is one of the few producers of rutile (titanium ore), which has become its most important export. The capture of the major mine by rebel forces, therefore, caused alarm.

Sierra Leone has natural wealth and the economic indications are encouraging. The currency is reasonably stable and inflation is falling. The government does appear to be working towards improvement by abolishing the corruption of the previous administration, and has clamped down on diamond smuggling. However, Sierra Leone is one of the poorest countries in the world and its people will become poorer while civil war rages.

THE REPUBLIC OF SIERRA LEONE	
STATUS	Republic
AREA	72,325 sq km (27,920 sq miles)
POPULATION	4,297,000
CAPITAL	Freetown
LANGUAGE	English, Krio Temne, Mende
RELIGION	52% animist, 39% Muslim, 8% Christian
CURRENCY	leone (SLL)
ORGANIZATIONS	Comm., ECOWAS, OAS, UN

PHYSICAL GEOGRAPHY

The Republic of Singapore comprises the main island of Singapore and 57 other islands, located at the southern tip of the Malay peninsula.

CLIMATE

The climate is equatorial, with high temperatures, humidity and heavy rainfall throughout the year.

POLITICS AND RECENT HISTORY

Singapore achieved independence from Britain in 1959, and in that year elections were won by the People's Action Party (PAP), led by Lee Kuan Yew. Singapore became a founder member of the Federation of Malaysia in 1963, but left only two years later and on 9 August 1965 was proclaimed a republic.

The PAP has remained in power since 1959, giving Singapore the appearance of a one-party state, especially as the opposition Singapore Democratic party is weak and divided. In elections held in 1988, the PAP was returned with more than 61 per cent of the vote. Lee Kuan Yew remained in power until 1990, when he retired in favour of

Goh Chok Tong. Two years later Goh was returned with 72 per cent of the vote.

August 1993 saw the first-ever direct presidential election in Singapore. Although the presidency is a non-party post, the election was won by Ong Teng Cheong of the PAP, who had to resign his post of deputy prime minister in order to stand. The government is, from time to time, accused of human rights abuses and recently earned the disapproval of western nations by hosting a visit from the head of the Myanmar's ruling council.

ECONOMY

Over the last three decades Singapore has undergone an astonishing process of econom-

THE REPUBLIC OF SINGAPORE	
STATUS	Republic
AREA	616 sq km (238 sq miles)
POPULATION	2,930,000
CAPITAL	Singapore
LANGUAGE	Malay, Chinese (Mandarin), Tamil, English
RELIGION	Daoist, Buddhist, Muslim, Christian, Hindu
CURRENCY	Singapore dollar (SGD)
ORGANIZATIONS	ASEAN, Col. Plan, Comm., UN

ic expansion. Gross domestic product (GDP) has grown consistently, with latest figures at about 10 per cent. The financial services sector is particularly buoyant and now contributes 13 per cent of GDP. Singapore has full employment and the highest rate of home ownership and national savings anywhere in the world. Foreign investment is so great that there is one foreign company in Singapore for every 1,000 people. Amid this prosperity, the government adopts a curiously secretive attitude to economic statistics, and in 1993 prosecuted five citizens for publishing GDP estimates.

The only significant economic problem is a chronic shortage of labour, and the government, although conscious of this, has resisted an influx of foreigners. The problem may be solved by technology. The entire island is to be equipped with an optical fibre cable network, reaching every home and office. No doubt this will further enhance Singapore's business activities.

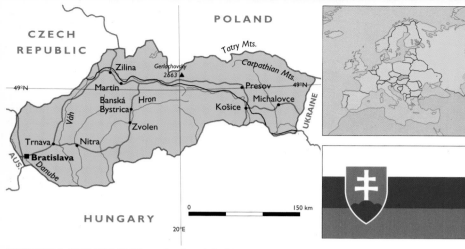

PHYSICAL GEOGRAPHY

Slovakia is mountainous towards its border with Poland in the north, with peaks in the Tatry Mountains exceeding 2,500 m (8,200 ft). The land slopes down to the lowland plains of the Danube.

CLIMATE

The climate is continental with warm summers and cold winters. Mountainous areas are snow covered for more than a third of the year.

POLITICS AND RECENT HISTORY

Czechoslovakia emerged from Soviet domination in 1989, under the leadership of Vaclav Havel. The new republic appeared to have a good chance of achieving prosperity under a democratic government, intent on developing closer ties with western Europe. However, the nationalist ambitions of the minority Slovak people in the east resulted in what has become known as the 'velvet divorce'. The Czech and Slovak Republics separated and on 1 January 1993 became independent states.

Slovakia was led by Prime Minister Vladimír Mečiar, the leader of the Movement for a Democratic Slovakia (HZDS), originally in coalition with the Slovakia National Party (SNS). The coalition survived until March 1994, when Mečiar's government was forced to resign following a vote of no confidence. An alliance of five parties nominated Jozef Moravcik as the new prime minister.

In October 1994, the first general election since Slovakia's independence was held. The HZDS emerged as the largest single party and Mečiar began his third term as prime minister, after forging an unlikely coalition with two minor parties, one with extreme nationalist views, the other a hard-line Communist party. This political combination has caused some dismay in other countries, although Mečiar has declared his intention to apply for membership of the European Union.

ECONOMY

Slovakia's industrial inheritance was large manufacturing complexes devoted particularly to the armaments industry and specializing in tanks, armoured personnel carriers and artillery. The end of the Cold War has seen a collapse in the demand for such equipment, with the result that Slovakia's industrial output has seriously declined. The economy has also suffered from the separation of the Czech and Slovak republics – exports to the Czech Republic have halved. The resulting harsh economic conditions have been worsened by an influx of refugees from Romania and former Yugoslavia.

The Moravcik administration had seemed to achieve some economic stability and had embarked on an ambitious privatization programme. However, the restoration of Mečiar may signal slower progress. He has, for example, appointed as head of the privatization programme a hard-line Communist, opposed to the sale of state assets.

THE SLOVAK REPUBLIC	
STATUS	Republic
AREA	49,035 sq km (18,932 sq miles)
POPULATION	5,353,000
CAPITAL	Bratislava
LANGUAGE	Slovak, Hungarian minority
RELIGION	Roman Catholic
CURRENCY	Slovak crown or koruna
ORGANIZATIONS	Council of Europe, UN

PHYSICAL GEOGRAPHY
Slovenia is a mountainous state in which the highest land in the Julian Alps, to the northwest of the country, reaches more than 2,500 m (8,200 ft) above sea-level. The Sava and Drava are the main river systems.

CLIMATE
The climate generally shows continental tendencies with warm summers and cold winters, when snow is plentiful in the mountains. The small coastal strip has a Mediterranean regime.

POLITICS AND RECENT HISTORY
In April 1990, the return of a six-party coalition in free multi-party elections paved the way for the declaration of independence from Yugoslavia in June 1991. Although the Belgrade government inter-vened and sent troops to Slovenia, the conflict was short lived. The troops withdrew, Yugoslavia accepted the loss of the republic and, after an agreed three-month moratori-um, the new state was recog-nized internationally.

Slovenia has been able to divorce itself from the tragic events elsewhere in former Yugoslavia, mainly because it does not have the same ethnic diversity as in other republics: some 90 per cent of the popula-tion is Slovene. A new constitu-tion was introduced in December 1991, and a year later Milan Kucan was elected president. The government is led by Prime Minister Janez Drnovsek of the Liberal Democrats, who heads an essentially centrist coalition with support from the Christian Democrats and the Associated List (the former Communists).

Slovenia has started negoti-ations with the European Union with a view to gaining associate status. This dialogue had been vetoed by Italy because of a dispute, now resolved, over compensation for land near Trieste that Italy was forced to surrender in 1944.

ECONOMY
Slovenia has emerged from Communist rule with a fair degree of success, although it has lost the majority of its mar-kets within former Yugoslavia and has had to re-orient most of its trade. The early years of independence were difficult, with high inflation and a decline in the standard of liv-ing. Recovery is now under way. The tolar is stable and the economy is growing. The priva-tization programme, slow to get under way, is now gather-ing pace.

The government is invest-ing in infrastructure and, in particular, in road and rail links to its Adriatic port at Koper, in expectation of trade from there into central Europe. The tourist industry is also expanding, as Slovenia's ancient spa towns are refurbished.

Privatization, a likely key to future prosperity, has so far proceeded rather slowly. The government forecast that by the end of 1993, 400 state enter-prises would have been trans-ferred to private ownership, and by the end of 1994 the fig-ure would be 1,600.

THE REPUBLIC OF SLOVENIA	
STATUS	Republic
AREA	20,250 sq km (7,815 sq miles)
POPULATION	1,988,000
CAPITAL	Ljubljana
LANGUAGE	Slovene
RELIGION	Roman Catholic
CURRENCY	Slovene tolar (SLT)
ORGANIZATIONS	Council of Europe, UN

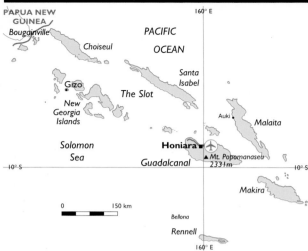

PHYSICAL GEOGRAPHY
The Solomon Islands archipelago contains several hundred islands of which six are large, volcanic, mountainous and forested. Guadalcanal, the dominant island, has the largest area of flat land.

CLIMATE
The climate in the northern islands of the group is hot and humid all year round, but a cool season develops further south. The wet season is between November and April.

POLITICS AND RECENT HISTORY
The British Solomon Islands protectorate became self-governing in 1976 and fully independent in 1978. It is a constitutional monarchy with the British monarch as head of state, represented by a Solomon Island governor-general who appoints the cabinet. Real executive power, however, rests with the prime minister, who recommends the members of the cabinet.

In the last general election held in May 1993, Francis Billy Hilly took the post of prime minister, leading an alliance of several political parties.

However, in late 1994 Hilly, having been dismissed by the Governor-General, resigned his post. Parliament appointed Solomon Mamaloni, prime minister on two previous occasions, in his place.

The Solomon Islands has an uneasy relationship with neighbouring Papua New Guinea because of the crisis in Bougainville. Papua New Guinea has accused the Solomon Islands of assisting the rebels, although the two countries have signed a peace agreement. As many as 5,000 Bougainvillians have sought refuge in the Solomons, adding to the demographic problems.

ECONOMY
The main products are palm oil, cocoa, fish, timber and wood products. There are some bauxite deposits and phosphates are mined on the island of Bellona. However, 90 per cent of the population is involved in subsistence agriculture, concentrated on the flat coastal zones. Development is restricted by the dense tropical forests which prohibit the growth of transport systems.

The economy has shown a decline since 1990. Fish, usually a prime export, has fallen sharply in value and there has also been a fall in cocoa and palm oil products. In contrast, exports of tropical logs have soared. The Solomon Islands is one of the few countries in the region still allowed to export logs. With increasing demand prices have risen, and the pace of timber extraction has accelerated, restriction rules apparently being disregarded by the logging companies. It has been reported that the entire population of one island is to be resettled to permit logging. The new prime minister, has promised to renew good logging practices but with an otherwise stagnant economy, a high birth rate and a mere 19 per cent of the adult population in work, he has little room for manoeuvre.

THE SOLOMON ISLANDS	
STATUS	Commonwealth Nation
AREA	29,790 sq km (11,500 sq miles)
POPULATION	355,000
CAPITAL	Honiara
LANGUAGE	English, pidgin English, native languages
RELIGION	95% Christian
CURRENCY	Solomon Islands dollar (SBD)
ORGANIZATIONS	Comm., UN

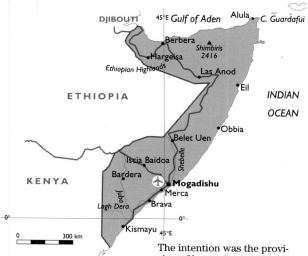

PHYSICAL GEOGRAPHY

To the north, an extension of the Ethiopian highlands provides rugged terrain. Southwards the flat landscape comprises scrub, although the coastal areas and valleys are more fertile.

CLIMATE

Temperatures approach 30°C (86°F) year-round. Inland and northern areas are relatively arid and the Indian Ocean coasts receive moderate rains.

POLITICS AND RECENT HISTORY

Somalia was formed in 1960 by the merger of the newly independent territories which had been the British Somaliland Protectorate and the Italian Trusteeship Territory of Somalia. In 1969 General Mohammed Siad Barre seized power in a coup. His military dictatorship survived until 1991, when he was overthrown. Thereafter effective government broke down, clan rivalries and armed conflicts erupted, populations were displaced and severe famine resulted.

By late 1992 the situation had deteriorated so far that the United Nations (UN), with US forces in a lead role, intervened.

The intention was the provision of humanitarian aid and the installation of a government of national reconciliation within 18 months. The UN mission failed. Fighting between UN forces, in which Pakistani and US soldiers were killed, and Somali warlords ensured that the supply of aid faltered and no government could be created. Domestic pressure persuaded President Clinton to withdraw US forces. France and Belgium followed suit and the last UN forces left Somalia in March 1995.

Control has passed to the two most powerful warlords. General Farah Aideed controls the south of the country while Ali Mahdi Muhammad rules the north. Despite all the difficulties, UN aid agencies returned in mid-1995 in an effort to bring

THE SOMALI DEMOCRATIC REPUBLIC	
STATUS	Republic
AREA	630,000 sq km (243,180 sq miles)
POPULATION	8,954,000
CAPITAL	Mogadishu (Muqdisho)
LANGUAGE	Somali, Arabic, English, Italian
RELIGION	Muslim majority, Roman Catholic minority
CURRENCY	Somaliland shilling (SOS)
ORGANIZATIONS	Arab League, OAU, UN

relief to the Somali people. Meanwhile, in the north, the territory which was once British Somaliland declared its independence from Somalia and named itself Somaliland. Although its president Mohammad Ibrahim Egal faces opposition from clan chiefs, he is gradually building an administrative structure in relative peace. He does so without international aid because his country is not recognised by the UN or any country.

ECONOMY

The lack of coherent government has meant the total collapse of the Somali economy. UN operations were able to ensure the delivery of aid and overcame the worst effects of the famine. Reasonable rainfall in recent years has meant that crops can be harvested. However, recovery is dependent upon stable government.

In President Egal's regime, however, the port of Berbera is flourishing with the export of livestock to the Gulf States and the import of goods destined for Ethiopia and Kenya. The territory has introduced its own new Somaliland shilling, which has been minted in Britain.

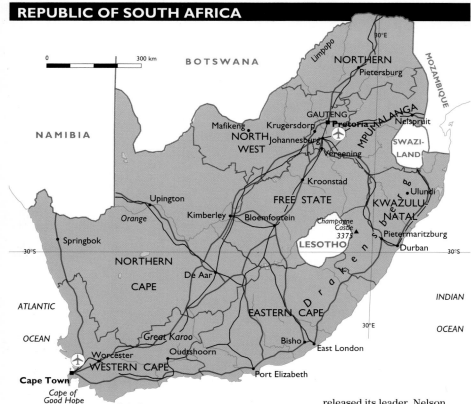

PHYSICAL GEOGRAPHY

Much of South Africa is ancient plateaux covered with grass-land or bush, depending on the degree of aridity, and drained in the west by the Orange river system and in the east by the Limpopo and its tributaries. Mountain ridges running east-wards from the Cape culminate in the Drakensberg mountain range, overlooking the coastal lowlands of Natal.

CLIMATE

There is a climatic contrast between the interior, which receives the majority of its rainfall in summer, and the coast around Cape Town, which experiences a Mediterranean style of climate with winter rains. Summers are warm and winters are mild everywhere.

POLITICS AND RECENT HISTORY

When F. W. de Klerk assumed the presidency in September 1989 he set about transforming the country's politics and abolishing apartheid. In February 1990 he removed the ban upon the African National Congress (ANC), and nine days later

THE REPUBLIC OF SOUTH AFRICA	
STATUS	Republic
AREA	1,220,845 sq km (471,369 sq miles)
POPULATION	40,435,000
CAPITAL	Pretoria (administrative) Cape Town (legislative)
LANGUAGE	Afrikaans, English, various African languages
RELIGION	Christian majority, Hindu, Jewish and Muslim minorities
CURRENCY	rand (ZAR)
ORGANIZATIONS	Commonwealth, OAU, SADC, UN

released its leader, Nelson Mandela, from detention. There followed several years of difficult negotiations between de Klerk's government and the ANC. Brutal conflict was waged between ANC supporters, who are mainly Xhosa, and the followers of the Inkatha Freedom Party, led by Chief Mangosutho Buthelezi.

Nevertheless, agreement was finally reached. The white population voted in favour of constitutional equality for all, parliament ratified a new constitution and elections were duly held in April 1994. The ANC gained a comfortable majority and formed a Government of National Unity, sharing power with F.W. de Klerk's National Party and the Inkatha Freedom Party. Nelson Mandela became president and de Klerk vice-president. Although there were some alle-

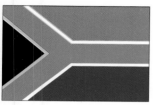

this and a more equitable distribution of benefits in a reasonable time-frame. Inevitably, the black population, many of whom are unemployed, will have expectations that the government will not be able to meet.

gations of fraud, the election was generally held to be fair.

ECONOMY

Agriculture is limited by poor soils but sheep and cattle are extensively grazed. The main crops are maize, wheat, sugarcane, vegetables, cotton and vines. Wine is an important export commodity.

South Africa abounds in minerals. Diamonds, gold, platinum, silver, uranium, copper, manganese and asbestos are mined and nearly 80 per cent of the continent's coal reserves are in South Africa. Manufacturing and engineering are concentrated in southern Transvaal and around the ports.

The South African economy is emerging from many years of international sanctions and more recently from the effects of recession. The economy is growing, foreigners are investing in the country and the traditional trading relationship with Britain is being revived, boosted by Queen Elizabeth II's visit in March 1995.

Undoubtedly, South Africa has the wealth and infrastructure to achieve prosperity. The challenge will be to achieve

EASTERN CAPE

STATUS	Province
AREA	174,405 sq km (67,338 sq miles)
POPULATION	5,900,000
CAPITAL	Bisho

FREE STATE

STATUS	Province
AREA	123,893 sq km (47,835 sq miles)
POPULATION	2,500,000
CAPITAL	Bloemfontein

GAUTENG

STATUS	Province
AREA	18,078 sq km (6,980 sq miles)
POPULATION	6,500,000
CAPITAL	Johannesburg

KWAZULU-NATAL

STATUS	Province
AREA	90,925 sq km (35,106 sq miles)
POPULATION	8,000,000
CAPITAL	Ulundi

MPUMALANGA

STATUS	Province
AREA	73,377 sq km (28,331 sq miles)
POPULATION	2,600,000
CAPITAL	Nelspruit

NORTHERN

STATUS	Province
AREA	121,766 sq km (47,014 sq miles)
POPULATION	4,700,000
CAPITAL	Pietersburg

NORTHERN CAPE

STATUS	Province
AREA	369,552 sq km (142,684 sq miles)
POPULATION	700,000
CAPITAL	Kimberley

NORTH WEST

STATUS	Province
AREA	120,170 sq km (46,398 sq miles)
POPULATION	3,300,000
CAPITAL	Mafikeng

WESTERN CAPE

STATUS	Province
AREA	128,679 sq km (49,683 sq miles)
POPULATION	3,400,000
CAPITAL	Cape Town

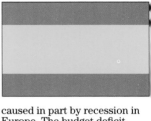

PHYSICAL GEOGRAPHY
Mainland Spain is mostly high plateaux. The principal mountain ranges are the Pyrenees and the central Sierras northwest of Madrid.

CLIMATE
Temperate maritime conditions in the north grade into hotter, drier regimes in the south. Winter can bring very cold weather to the central plateau.

POLITICS AND RECENT HISTORY
In 1975 King Juan Carlos I succeeded to the monarchy upon the death of the fascist dictator General Franco, and was instrumental in restoring democracy and the defeat of an attempted coup in 1981. In 1982 Spain was admitted to the North Atlantic Treaty Organization and in 1986 became a full member of the European Union.

In the June 1993 general elections, the Socialist Party, under Prime Minister Felipe González Márquez, was returned for a fourth term, defeating the conservative Popular Party led by José Maria Aznar. However, González was unable to secure an overall majority, and although neither of the minority Catalan and Basque parties would agree to

form a coalition, they undertook to support him in order to ensure economic progress.

After the 1993 election, González lost support following allegations that his government waged a dirty war against Basque separatists during the 1980s, and due to the nation's economic difficulties. Then in the March 1996 general election the Popular Party under Aznar narrowly defeated the Socialists.

Although the threat from the Basque separatist group ETA continues with periodic terrorist attacks, there is speculation that they may abandon violence.

ECONOMY
The Spanish government has been forced to tackle quite serious economic difficulties

THE KINGDOM OF SPAIN	
STATUS	Kingdom
AREA	504,880 sq km (194,885 sq miles)
POPULATION	39,167,000
CAPITAL	Madrid
LANGUAGE	Spanish (Castilian), Catalan, Basque, Galician
RELIGION	Roman Catholic
CURRENCY	Spanish peseta
ORGANIZATIONS	Council of Europe, EEA, EU, NATO, OECD, UN, WEU

caused in part by recession in Europe. The budget deficit, estimated to have risen to 6 per cent of gross domestic product by the end of 1993, has been decreased by austere measures, including freezing public sector pay, increasing fuel taxes and cutting health spending, all of which provoked trade union opposition. Unemployment has fallen slightly, but remains high. Fortunately, the tourist industry has experienced a huge upswing in bookings, prompted by the falling popularity of Florida.

During its 11 years of office, the González government moved away from its socialist origins in terms of economic policy and has embarked on a major privatization programme. The Socialist government was particularly successful in attracting money from overseas – Spain is the fourth largest recipient of direct foreign investment, although this has fallen, largely because of the instability of the peseta.

The fishing industry is of great economic importance. Spanish fishermen have been accused of exceeding quotas and using illegal nets, giving rise to conflict with other countries, notably Britain and Canada.

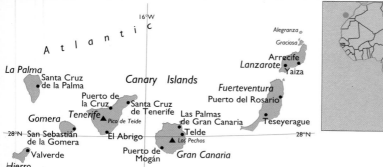

CANARY ISLANDS		
STATUS......**Island Provinces of Spain**		
AREA.........................**7,275 sq km** **(2,810 sq miles)**		
POPULATION.................**1,493,784**		
CAPITAL....**Las Palmas (Gran Canaria)** **Santa-Cruz (Tenerife)**		

CEUTA		
STATUS......**Spanish External Province**		
AREA...........................**19.5 sq km** **(7.5 sq miles)**		
POPULATION.....................**67,615**		
CAPITAL.............................**Ceuta**		

MELILLA		
STATUS......**Spanish External Province**		
AREA.............................**13 sq km** **(5 sq miles)**		
POPULATION.....................**56,600**		
CAPITAL.............................**Melilla**		

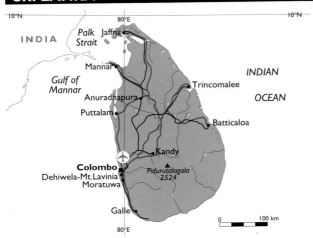

PHYSICAL GEOGRAPHY

The island of Sri Lanka varies in character from tropical jungle to lush temperate hills. The highest point is Mt. Pidurutalagala, which reaches 2,524 m (8,280 ft) above sealevel.

CLIMATE

Sri Lanka lies just north of the Equator, and its climate varies from exceptionally hot in northern regions, with temperatures over 38°C (100°F), to the cooler south, tempered by breezes. The two monsoon periods centre on May and November.

POLITICS AND RECENT HISTORY

Colonized in turn by the Portuguese, the Dutch and the British, Sri Lanka has a record of impartial democratic elections. The country became independent in 1948 and in 1972 became a republic, changing its name from Ceylon to the Republic of Sri Lanka.

For nearly two decades until 1977, the socialist prime minister Mrs Sirima Bandaranaike attempted to unite the island under a single national language, Sinhalese,

but this antagonized Tamils and Catholics alike. Since then, a system of government with an executive president has been installed and the United National Party (UNP) held power until parliamentary elections in August 1994, when Chandrika Kumaratunga of the People's Alliance, and daughter of Bandaranaike, was victorious. Three months later, Kumaratunga won the presidential election against Dissanayake, the widow of the UNP candidate who had been assassinated two weeks previously. Kumaratunga appointed her mother prime minister. Kumaratunga, who enjoys the support of many Tamils, made determined efforts to bring an end to the civil war. These have not been successful.

ECONOMY

Tea, coffee and gemstones have formed the basis of Sri Lanka's economy for generations, and the island is soon expected to regain its position as the largest producer of tea in the world.

However, in the last five years the manufacturing sector has overtaken tea and other crops as the main export earner. This sector, in particular the manufacture of clothing and textiles, now accounts for 65 per cent of total exports.

Tourism, seriously affected by the separatist activities during the 1980s, is re-emerging as a vital industry.

The growth-rate of the economy is increasing, and the budget introduced by the new government is clearly aimed at carrying forward the liberalization policies introduced by its predecessor. This is likely to include privatization of state-owned assets.

THE DEMOCRATIC SOCIALIST REPUBLIC OF SRI LANKA	
STATUS	Republic
AREA	65,610 sq km (25,325 sq miles)
POPULATION	17,619,000
CAPITAL	Colombo
LANGUAGE	Sinhala, Tamil, English
RELIGION	70% Buddhist, 15% Hindu, Roman Catholic and Muslim minorities
CURRENCY	Sri Lanka rupee (LKR)
ORGANIZATIONS	Col. Plan, Comm., UN

PHYSICAL GEOGRAPHY
Sudan lies within the basin of the Upper Nile river, much of which is arid plain, although marshy conditions are found in the south. The only areas of high relief are along the Red Sea coast and at the southern boundary of the country.

CLIMATE
The Sudanese climate is hot and arid, with summer temperatures reaching 40°C (104°F) and light rains in summer.

POLITICS AND RECENT HISTORY
President Omar al Bashir came to power in 1989 when he led a successful army coup against the elected coalition government of Sadig al Mahdi. He has instituted a revolutionary council over which he presides, and his government claims to adhere to Muslim fundamentalist principles. In reality, it is a corrupt and brutal regime that commands little support and controls the country with a large security force.

The government has pursued a military campaign against the Christian and animist populations of southern Sudan. The coup which brought the president to power ended the tentative moves towards peace that his predecessor had pursued. The war against the Sudan People's Liberation Army (SPLA) has cost thousands of lives, and an estimated two million people have been driven from their homes, causing massive refugee problems along southern Sudanese borders and in adjoining countries. A ceasefire brought about by the mediation of ex-US president Jimmy Carter may bring some relief.

The US government has declared the state to be a supporter of terrorism, and at the end of 1993 relations with the UK reached a new low. The UK ambassador was asked to leave Sudan, following a refusal by the Archbishop of Canterbury, who was visiting south Sudan at the invitation of local churches, to visit Khartoum as a guest of the government.

ECONOMY
The Sudanese economy is agricultural and highly dependent on the River Nile to provide irrigation for the growing of cereals, groundnuts, sugar cane and cotton. However, livestock rearing, often on a nomadic basis, is the main occupation. In southern Sudan there is no effective economy. The challenge is for aid agencies to feed the starving thousands in the refugee camps.

The government has been forced to re-introduce price controls in an effort to control inflation which has reached 100 per cent. Prices had soared following an economic liberalization programme started in 1990 under pressure from the International Monetary Fund (IMF). Since then relations with the IMF have deteriorated over the issue of a $1.5-million debt. International aid has in any case virtually ceased, apart from the efforts of relief agencies in the south.

THE REPUBLIC OF SUDAN	
STATUS	Republic
AREA	2,505,815 sq km (967,245 sq miles)
POPULATION	24,941,000
CAPITAL	Khartoum
LANGUAGE	Arabic, tribal languages
RELIGION	60% Sunni Muslim, animist and Christian
CURRENCY	Sudanese pound (SDP)
ORGANIZATIONS	Arab League, OAU, UN

SURINAM

PHYSICAL GEOGRAPHY
Most of Surinam is covered by tropical forests. There is a fertile coastal plain, central plateaux and the the Guiana Highlands.

CLIMATE
The climate is tropical and hot, with high rainfall and humidity.

POLITICS AND RECENT HISTORY
Formerly Dutch Guiana , Surinam became fully independent in 1975. For most of the time until 1991, following coups in 1980 and 1990, the government was under military control. Elections in 1991, however, established a civilian coalition government under the leadership of President Runaldo Venetiaan, who amended the constitution to limit the power of the military.

In 1992 a peace agreement was reached between guerrillas and government troops, ending a long-lasting virtual civil war.

ECONOMY
The main exported products of Surinam are bauxite, alumina and aluminium, although shrimp exports are growing and the timber industry has potential for growth. Agricultural exports include bananas, coconuts and citrus fruits.

Surinam has been accustomed to income from massive aid packages from the Netherlands. At various times when aid has been suspended, such as following the various coups, the economy has rapidly worsened.

THE REPUBLIC OF SURINAM	
STATUS	Republic
AREA	163,820 sq km (63,235 sq miles)
POPULATION	414,000
CAPITAL	Paramaribo
LANGUAGE	Dutch, English, Spanish, Surinamese (Sranang Tongo), Hindi
RELIGION	45% Christian, 28% Hindu, 20% Muslim
CURRENCY	Surinam guilder (SRG)
ORGANIZATIONS	OAS, UN

SWAZILAND

PHYSICAL GEOGRAPHY
Plateaux descend eastwards from heights of over 1,800 m (5,900 ft) to 300 m (1,000 ft).

CLIMATE
The climate varies with altitude. A warm wet season lasts from October to March, while May to December are dry and cooler.

POLITICS AND RECENT HISTORY
Swaziland gained independence from Britain in 1968. In 1976 the constitution was suspended, and a new constitution followed in which the monarch assumed supreme power. The monarchy has been under threat from dissident members of the royal family and politicians, but all have apparently been quashed with ease.

ECONOMY
Agriculture involves over three quarters of the population, generally as subsistence farmers. Cash crops include sugar, citrus fruits and pineapples. Asbestos and coal are important, and industries include sugar refining and wood pulping. The tourist trade, consisting mainly of visitors from South Africa, is increasing.

The end of the civil war in Mozambique may herald improved economic conditions and Swaziland plans to renew the routing of its exports through Mozambique. There is, however, a shortage of foreign capital and a population swollen by refugees from Mozambique. Swaziland is a member of the South African Customs Union and uses the South African rand as well as its own currency. Its economy is, therefore, closely associated with South African affairs.

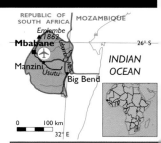

THE KINGDOM OF SWAZILAND	
STATUS	Kingdom
AREA	17,365 sq km (6,705 sq miles)
POPULATION	823,000
CAPITAL	Mbabane
LANGUAGE	English, Siswati
RELIGION	60% Christian, 40% traditional beliefs
CURRENCY	lilangeni (SZL), South African rand (ZAR)
ORGANIZATIONS	Comm., OAU, SADC, UN

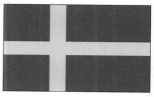

PHYSICAL GEOGRAPHY

Forested mountains cover the northern half of the country. To the south is a region characterized by thousands of lakes, then a southern upland and in the far south the fertile plain of Scania.

CLIMATE

The climate varies with latitudinal extent: winters are long in the north, and heavy snow can persist for between four and seven months. The rainfall, distributed all year round, is heavy but decreases eastwards.

POLITICS AND RECENT HISTORY

Sweden is a constitutional monarchy in which the King is head of state but has no executive power.

In September 1994, the Conservative led coalition under Carl Bildt was defeated at a general election. The Social Democrats, traditionally the dominant party in Swedish politics gained most seats but fell short of an overall majority. Their leader, Ingvar Carlsson, formed a minority government, relying on support from the Greens, former Communists and the Centre Party.

A formal application for membership of the European Union (EU) was submitted in 1991. In March 1994 Sweden and the EU agreed on membership terms for entrance to the EU from 1995. A referendum on the subject was held in November 1994. There had been considerable doubt as to the outcome, but the majority voted in favour of EU membership.

ECONOMY

Sweden's wealth in timber, hydro-electric power and minerals, in particular iron ore, copper and uranium, has been the basis of a prosperous manufacturing economy.

Until the late 1980s there was almost no unemployment, but rates have climbed alarmingly in recent years, although latest figures show marginal improvement. The new government has not only to tackle this problem, but that of a budget deficit. The root cause is Sweden's extremely generous welfare system, which is in need of overhaul. So far, the government has declared a budget which cuts public spending and increases taxation.

Sweden has a powerful environmentalist lobby and its population has voted for the dismantling of all nuclear reactors by the year 2010. Environmental concerns also delayed agreement with the Danish government on a bridge between Malmö and Copenhagen, but agreement has now been reached.

THE KINGDOM OF SWEDEN	
STATUS	Kingdom
AREA	449,790 sq km (173,620 sq miles)
POPULATION	8,801,000
CAPITAL	Stockholm
LANGUAGE	Swedish, Finnish, Lappish
RELIGION	95% Evangelical Lutheran
CURRENCY	Swedish krona (SEK)
ORGANIZATIONS	Council of Europe, EEA, EU, OECD, UN

PHYSICAL GEOGRAPHY

Switzerland is the most mountainous nation in Europe. The southern half of the country lies within the Alps while its northwestern border with France coincides with the Jura range. The remainder of the country is high plateau.

CLIMATE

Because of its altitude Switzerland experiences cold winters with heavy snowfalls. Summers are warm.

POLITICS AND RECENT HISTORY

Switzerland is a federal republic with a government formed from a broad coalition of the major political parties. The cabinet, known as the Federal Council, is elected by the Assembly, and it in turn appoints one of its seven members to serve for a year as president, although the post is effectively that of prime minister. The unusual feature of Swiss politics is its system of direct democracy. The population is called upon regularly to vote upon a variety of issues, sometimes major and sometimes parochial. Referenda are obligatory if supported by a required number of voters, the number varying depending upon the nature of the topic. This system is claimed to be particularly appropriate to Switzerland,

with its linguistic diversity.

Switzerland has maintained a policy of very strict political neutrality for many years. It does not even belong to the United Nations, although it does now appoint an observer. Its neutral stance has traditionally meant high levels of defence spending, which have come under scrutiny in recent years.

ECONOMY

Agriculture is based mainly on dairy farming. Major crops include hay, wheat, barley and potatoes. Industry plays a major role in Switzerland's economy, centred on metal engineering, watchmaking, food processing, textiles and chemicals. Tourism is an important source of income and employment, and the financial services sector, especially banking, is also of great importance.

THE SWISS CONFEDERATION	
STATUS	Federation
AREA	41,285 sq km (15,935 sq miles)
POPULATION	7,005,000
CAPITAL	Bern (Berne)
LANGUAGE	German, French, Italian, Romansch
RELIGION	48% Roman Catholic, 44% Protestant, Jewish minority
CURRENCY	Swiss franc (CHF)
ORGANIZATIONS	Council of Europe, EFTA, OECD, UN observer post

Although the Swiss economy suffered from the effects of recession, it has emerged in a sound condition and industry is doing well. The Swiss franc has continued to maintain its traditional strength.

However, economic reform is known to be necessary. Much of Swiss industry operates as cartels, maintaining high prices, and this position is not indefinitely sustainable. The model may be the watch industry, which has successfully regenerated following the blow of Japanese competition in the 1970s. Switzerland has opted not to join the European Union (EU) but is looking increasingly isolated since Austria's membership. There is a discernable move in public opinion toward EU membership but still a strong sense of independence. Switzerland's population recently caused some irritation to both its own government and the EU when it voted not to permit the passage of foreign trucks carrying freight across its territory.

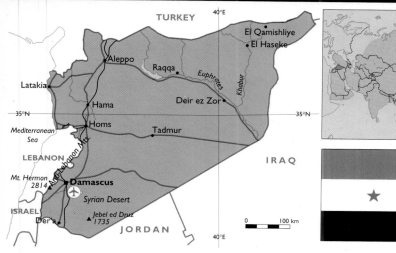

PHYSICAL GEOGRAPHY
In the north of the country, the coastal plain is backed by a low range of hills. The interior is a vast plateau through which the River Euphrates has cut its valley. In the south, the Anti-Lebanon range lies along the border with Lebanon, inland of which is the Syrian desert.

CLIMATE
In coastal regions, Mediterranean conditions with warm summers and mild moist winters prevail. The climate is hotter and drier inland.

POLITICS AND RECENT HISTORY
Syria is ruled by the Arab Socialist Renaissance (*Ba'ath*) Party. Its leader, General Hafez al Assad, was re-elected unopposed in 1991 for a fourth term as president and wields considerable authority. The country, regarded by western nations as a pariah for its alleged support of terrorism, gained favour by joining the coalition of allied forces in the Gulf War. Its military intervention in Lebanon has brought a degree of peace and stability to that country.

The relationship with Israel is the dominant issue for the Syrian government. The reaction to the Palestine Liberation Organization–Israeli agreement was cool because Yasser Arafat had failed to co-ordinate with Arab leaders, and President Assad has shown some support for Palestinian factions that oppose the deal. After prolonged manoeuvrings by both sides, agreement has been reached on the framework for peace talks to be held in New York. There is no doubt that Syria will accept nothing less than complete Israeli withdrawal from the Golan Heights, which have been occupied since 1967.

ECONOMY
The Syrian economy suffered from the break-up of the Soviet Union, far and away its most important trading partner. This event prompted the government to introduce economic reform, mainly to encourage investment and to stop Syrian business people removing wealth from the country. The result is that the economy is showing something of a boom, although according to the International Monetary Fund further action is necessary – bureaucracy within the public institutions is unchanged.

Agriculture, which provides 25 per cent of gross domestic product, is also flourishing. Cotton and cereals are the major products. Oil production and export have sharply increased in volume. Syria's participation in the Gulf War has encouraged oil-rich states, especially Kuwait, to provide substantial loans on preferential terms to invest in power generation and telecommunications. This development of infrastructure is necessary, but the greater share of the Syrian budget may continue to go towards its military activities.

THE SYRIAN ARAB REPUBLIC	
STATUS	Republic
AREA	185,680 sq km (71,675 sq miles)
POPULATION	13,844,000
CAPITAL	Damascus (Dimashq) (Esh Sham)
LANGUAGE	Arabic
RELIGION	65% Sunni Muslim, Shi'a Muslim and Christian minorities
CURRENCY	Syrian pound (SYP)
ORGANIZATIONS	Arab League, UN

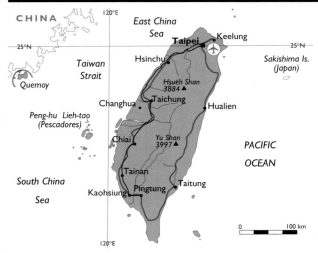

PHYSICAL GEOGRAPHY
Taiwan is a mountainous island with peaks approaching 4,000 m (13,100 ft). Coastal plains in the west are the focus of cultural and economic activity.

CLIMATE
Warm, humid climatic conditions prevail over much of the island, although snow falls in the mountains in winter. The monsoon rains fall between June to August, with an annual average rainfall of 2,600 mm (102 inches).

POLITICS AND RECENT HISTORY
The ruling Kuomintang Party (KMT) has governed Taiwan since 1949 when its founders fled Mao Tse Tung's Communism. Opposition parties have been allowed since 1987.

For many years the KMT government maintained that it represented all China. This stance quite obviously lacked credibility and the opposition Democratic Progressive Party, responding to this view, has announced its intention to declare formal independence. The ruling KMT under President Lee Teng-Hui, although now far from its origi-nal position, shrinks from this step for fear of encouraging Chinese invasion.

Dialogue between China and Taiwan has recently started. In August 1994, the two nations were able to agree on some issues, including the return of illegal immigrants. Nevertheless, China is clearly not intending to alter her attitude towards Taiwan in any fundamental way. Chinese protests were loud when the US allowed President Lee to visit in 1995 and China will surely block Taiwanese efforts to recover its seat at the United Nations.

ECONOMICS
In contrast to its Chinese neighbour, Taiwan has enjoyed prosperity with high incomes and a healthy trade surplus. This wealth is based on a manufacturing industry originally founded on textiles, but more recently on high-tech electronic goods.

As much as 40 per cent of the economy, however, is reckoned to be in illegal trading, and a black market lending system supports many small to medium industrial concerns. Agriculture is highly productive, with rice, sugar, tea and fruit as the leading crops.

The economy suffers from the drainage of its wealth as a result of investment in China. This is, however, partly offset by the sudden dramatic increase in trade between the two countries, despite an official Chinese boycott.

REPUBLIC OF CHINA ON TAIWAN	
STATUS	Island 'Republic of China'
AREA	35,990 sq km (13,890 sq miles)
POPULATION	20,600,000
CAPITAL	Taipei (T'ai-pei)
LANGUAGE	Mandarin Chinese, Taiwanese
RELIGION	Buddhist majority. Muslim, Daoist and Christian minorities
CURRENCY	New Taiwan dollar (TWD), yuan (CNY)
ORGANIZATIONS	none

PHYSICAL GEOGRAPHY

Tajikistan, with the western end of the Tien Shan and part of the Pamir range within its borders, is extremely rugged and mountainous. The only significant lowland lies within the Fergana valley.

CLIMATE

In the western areas of less severe altitude summers are warm but winters are generally cold.

POLITICS AND RECENT HISTORY

Tajikistan was internationally recognized as an independent republic upon the dissolution of the Soviet Union in 1991, and it became a member of the Commonwealth of Independent States. Although originally a presidential republic, in late 1993 parliament voted to abolish the presidency and create a parliamentary republic. Abdulmalik Abdulladzhanov became prime minister.

He was, however, forced to resign because of the worsening economic situation and the was replaced by Imomali Rakhmonov. In constitutional elections held in November 1994, the presidency was restored and Rakhmonov took that office.

For most of its short period of independence, Tajikistan has been plagued by a brutal civil war which has claimed the lives of thousands. The government portrays the rebels as Islamic fundamentalists, but the conflict is in reality between rival clans, whose feuding was contained during the Soviet era.

Russia has provided support to the undemocratic Tajik government and has stationed troops in the country, in an effort to keep the peace. However, because the Russian forces have sustained losses, President Yeltsin has urged the government to negotiate peace with the rebels.

ECONOMY

Tajikistan was never a wealthy republic, but it survived within the Soviet Union by exchanging its agricultural produce, mainly cotton, for energy supplies and consumer goods. Civil war has now brought the fragile economy near to collapse; even agriculture has deteriorated as facilities have been destroyed and communities driven from the land. Food aid has been needed for the starving population. Government workers are seldom paid.

Industrial production, of which aluminium smelting is the main component, has fallen drastically. Inflation is rife although the government has risked introducing its own new currency (also called the rouble) in the hope of achieving some stability.

If the civil war can be brought to an end, there are hopes for eventual prosperity. Tajikistan has important silver reserves and its gold mines have attracted foreign investment despite the internal conflict.

THE REPUBLIC OF TAJIKISTAN	
STATUS	Republic
AREA	143,100 sq km (55,235 sq miles)
POPULATION	5,638,000
CAPITAL	Dushanbe
LANGUAGE	Tajik, Uzbek, Russian
RELIGION	Sunni Muslim
CURRENCY	Tajikistan rouble
ORGANIZATIONS	CIS, UN

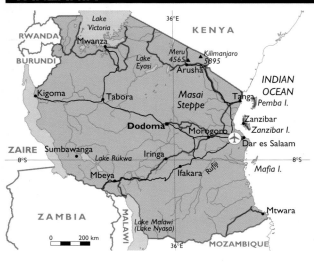

PHYSICAL GEOGRAPHY

Apart from a forested coastal plain, most of mainland Tanzania is a plateau at a height above 1,000 m (3,300 ft), with savannah vegetation, lying to the east of the Great Rift Valley. In the north Mt. Kilimanjaro dominates the landscape.

CLIMATE

The climate is tropical. The central lowlands and Zanzibar are particularly hot and humid. Temperatures inland average 25°C (77°F) throughout the year, and the heaviest rains fall from March to May.

POLITICS AND RECENT HISTORY

Tanganyika achieved independence from the United Kingdom in 1961 and Zanzibar, which includes the island of Pemba, did likewise in 1963. The two territories united to form Tanzania in 1964. Until 1992, Tanzania was a one-party Socialist state, but in May of that year a new law introduced multi-party democracy. The Socialist party nevertheless exercises considerable authority. The president, Ali Hassan Mwinyi, was first elected to office in 1985 following the retirement of Julius Nyerere, the country's leader since Tanganyikan independence. However, Mwinyi was unable to exercise full authority while Nyerere remained party chairman. The latter retired from this post in 1990, when Mwinyi was elected for a further five-year term.

President Mwinyi is expected to face a stiff challenge at the next elections – he sacked Augustine Mrema, Home Affairs Minister, who promptly joined the opposition party and declared his candidacy. Mrema commands popularity because of his consistent campaign against corruption, and may well be able to overturn Socialist rule.

ECONOMY

The Tanzanian economy is heavily based on agriculture, and most of the population is engaged in subsistence farming. Cash crops include coffee, tea, cotton, tobacco and sisal. Limited industrial activity includes the manufacture of textiles and cigarettes and some food processing.

The country is dependent upon assistance from the World Bank and the International Monetary Fund, but under President Mwinyi the government has been making efforts to liberalize and modernize the economy by loosening the centralist control of the socialist system.

The main scope for expansion is probably within the tourist industry. The potential afforded by wildlife and tropical beaches is not exploited to the same degree as in Kenya to the north. There may also be opportunities to increase revenue from minerals; diamonds are mined and there are small-scale workings of gold and gemstones. Exploration for oil and gas offshore has so far not met with success. New trade links with South Africa, impossible until recently, may help to reduce the national $1-billion trade deficit.

THE UNITED REPUBLIC OF TANZANIA	
STATUS	Republic
AREA	939,760 sq km (362,750 sq miles)
POPULATION	28,019,000
CAPITAL	Dodoma
LANGUAGE	Swahili, English
RELIGION	40% Christian, 35% Muslim
CURRENCY	Tanzanian shilling (TZS)
ORGANIZATIONS	Comm., OAU, SADC, UN

PHYSICAL GEOGRAPHY

The most populated centre of Thailand is an undulating plain through which the Chao Phraya flows. One third of the country is occupied by a plateau in the northeast, drained by tributaries of the Mekong river.

CLIMATE

The climate is tropical with three seasons. November to February is cool, March to May is hot and the monsoon rains arrive between May and October.

POLITICS AND RECENT HISTORY

Thailand, known as Siam before 1939, was an absolute monarchy until 1932 when a constitutional monarchy was established. For half a century Thai politics have seen a succession of civilian governments and military intervention. The King of Thailand, Bhumibol Adulyadej (Rama IX), is however regarded by the population as semi-divine. Following a period of military rule, constitutional government returned in September 1992. Chuan Leekpai, leader of the Democratic Party, won the election and became prime minister in a five-party coalition. This government lasted longer than most, but in May 1995, one party deserted the coalition and a confidence vote was lost. The prime minister promptly dissolved parliament and called a general election.

ECONOMY

Thailand is the world's biggest exporter of rice and rubber and also exploits tin, wolfram and natural gas. Manufacturing industry has overtaken agriculture in importance and textiles and clothing are the most important elements.

Tourism grew dramatically during the 1980s, and although it has levelled out since, is still important to the economy.

More recent ventures include phosphates, which have yet to be fully developed, and the shrimp industry, where Thailand is the world's biggest exporter.

Chuan Leekpai's government deserves credit for maintaining growth, for financial liberalization, for land reform and for channelling investment into infrastructure hitherto so poor as to hinder progress.

One major project will also benefit trade with Laos. The Friendship Bridge, linking the Laotian capital Vientiane with Thailand across the Mekong River, was opened in 1994.

THE KINGDOM OF THAILAND	
STATUS	Kingdom
AREA	514,000 sq km (198,405 sq miles)
POPULATION	58,584,000
CAPITAL	Bangkok (Krung Thep)
LANGUAGE	Thai
RELIGION	Buddhist, 4% Muslim
CURRENCY	baht (THB)
ORGANIZATIONS	ASEAN, Col. Plan, UN

TOGO

PHYSICAL GEOGRAPHY
Togo consists of plateaux rising to mountainous areas.

CLIMATE
The climate is tropical with rainfall of 890 mm (35 inches), heaviest to the west.

POLITICS AND RECENT HISTORY
Togo, as part of Togoland, was administered by France under UN Trusteeship until independence in 1960. In 1991 pressure from foreign creditors and internal strikes led to the transference of much of the president's power to a prime minister. Two years of increasing violence between the president's army and an interim government followed. In 1993, the first multiparty elections for 26 years took place. Due to massive rigging of the voting lists, the main opposition withdrew. President Eyadéma claimed victory with 96 per cent of the vote. Elections in early 1994 were won by opponents of the president, who had then to appoint a prime minister from the rival ranks.

ECONOMY
High-grade phosphates contribute half of the export revenue. Other exports include coffee, cocoa and cotton. The country is almost self-sufficient in food, and farming is the livelihood of most of the population. Industry is minimal. Oil is refined at Lomé and there is food processing and consumer goods manufacturing for the domestic market. Assistance from the International Monetary Fund is granted on condition that economic austerity measures are followed.

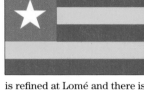

THE REPUBLIC OF TOGO	
STATUS	Republic
AREA	56,785 sq km (21,920 sq miles)
POPULATION	3,885,000
CAPITAL	Lomé
LANGUAGE	French, Kabre, Ewe
RELIGION	50% animist, 35% Christian, 15% Muslim
CURRENCY	CFA franc (W Africa) (XOF)
ORGANIZATIONS	ECOWAS, OAU, UN

TONGA

PHYSICAL GEOGRAPHY
Tonga is composed of some 170 islands, mostly uninhabited. About 60 per cent of the population live on Tongatapu.

CLIMATE
Temperatures are warm all year. Rainfall is plentiful, heaviest in February and March.

POLITICS AND RECENT HISTORY
Tonga is a hereditary monarchy in which the monarch holds significant executive power. The current ruler, King Taufa'ahau Tupou IV, succeeded his mother Queen Salote in 1965 and is the latest in a dynasty that has lasted 1,000 years. The nation became an independent republic within the Commonwealth in 1970; previously it had been a British protected state.

Recently a campaign for greater democracy has developed and in 1994, Tonga's first political party, the People's Party, was established by commoner M.P. Pohiva, to press for democratic reform.

ECONOMY
The basis of the economy is agricultural self-sufficiency in which land, rented from the government, is used to grow crops such as yams and manioc. Fishing is important in sustaining the population. The main export crops are bananas and products of the coconut palm.

Tourism has increased in recent years and revenue is accrued from a fisheries surveillance agreement with Tuvalu. The government has agreed with Intelsat to provide facilities for satellite communications and earns revenue by leasing these facilities to foreign operators.

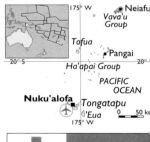

THE KINGDOM OF TONGA	
STATUS	Kingdom
AREA	699 sq km (270 sq miles)
POPULATION	98,000
CAPITAL	Nuku'alofa
LANGUAGE	Tongan, English
RELIGION	Christian
CURRENCY	pa'anga (TOP)
ORGANIZATIONS	Comm.

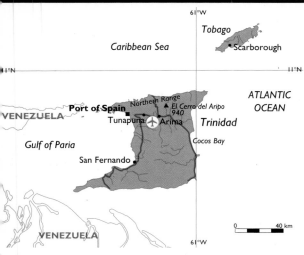

PHYSICAL GEOGRAPHY

The mountains of the Northern Range of Trinidad, which reach 940 m (3,085 ft) at Mt. Aripo, overlook undulating land and a flat central area, while swamps border Cocos Bay in the east. Tobago is mostly mountainous.

CLIMATE

High temperatures throughout the year vary little from 26°C (79°F). The annual rainfall is 1,631 mm (65 inches), falling most frequently between July and December.

POLITICS AND RECENT HISTORY

The two islands were combined as a British colony in 1899, achieving independence in 1962 and becoming a republic in 1976, remaining within the Commonwealth. The president is head of state, and executive power lies with the prime minister and cabinet.

Politics have been dominated by the People's National Movement (PNM), which governed for 30 years until 1986. In 1991 it was re-elected, heavily defeating a coalition (the National Alliance for Reconstruction) which had ruled from 1986 to 1991. The coalition had lost votes through unpopular austerity measures, a high incidence of unemployment and splits among its members. An attempted coup by Muslims in 1990 also shook the stability of the republic. The present prime minister and leader of the PNM is Patrick Manning, and the official opposition, a party derived from part of the coalition, is the United National Congress led by Baideo Panday. Tobago, with a parliament of its own since 1980, was granted full self-government in 1987.

Trinidad and Tobago participates actively in international organizations and has signed over 100 treaties. It is particularly active within the Caribbean Community and Common Market and the Organization of American States, advocating stronger political and economic co-operation within the region.

ECONOMY

The country's economy, historically based on sugar plantations, was transformed in the early 1970s when oil was discovered. Production has declined, and in the mid-1990s it was estimated that, at the present extraction rates, the proven reserves would last less than ten years. Oil now comes chiefly from offshore fields. A petrochemical industry is based on significant gas reserves which are calculated to last for 70 years. Oil and petrochemical products account for over 70 per cent of exports. Asphalt is also important. In the agricultural sector sugar, coffee, cocoa, citrus fruits and rubber are produced.

With the worldwide recession, Trinidad and Tobago has recorded a reversed migration trend as disillusioned nationals return home, and there has recently been a surge in land deals for retirement homes on Tobago.

THE REPUBLIC OF TRINIDAD AND TOBAGO	
STATUS	Republic
AREA	5,130 sq km (1,980 sq miles)
POPULATION	1,260,000
CAPITAL	Port of Spain
LANGUAGE	English, Hindi, French, Spanish
RELIGION	60% Christian, 25% Hindu, 6% Muslim
CURRENCY	Trinidad and Tobago dollar (TTD)
ORGANIZATIONS	Caricom, Comm., OAS, UN

TUNISIA

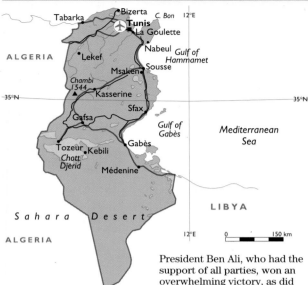

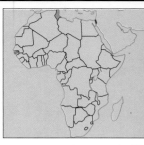

PHYSICAL GEOGRAPHY
The northern half of the country has rugged mountains, often tree-covered. These are separated from Saharan plains by the Chott Djerid, a low-lying area of salt pans.

CLIMATE
Summers are warm and winters mild, with temperatures ranging between 10–27°C (50–81°F). Northern parts experience winter rainfall, but arid conditions prevail in the south.

POLITICS AND RECENT HISTORY
In 1987 President Zine el-Abidine Ben Ali replaced the ailing Habib Bourguiba, who had held power since independence from France in 1957. The ruling party, the *Rassemblement Constitutionnel Démocratique* (RCD), and its allies hold all the seats in parliament. Prior to elections held in 1994, opposition parties had boycotted the polls, but following a constitutional change which guaranteed representation in parliament to opposition parties, the boycott was lifted. Despite the boycott,

President Ben Ali, who had the support of all parties, won an overwhelming victory, as did the RCD.

Tunisia enjoys a stable government but is nervous of the unpredictability of the regime in Libya to the east and the rise of Islamic fundamentalism in Algeria to the west. The Tunisian fundamentalist party is banned and the government clamps down firmly on its activities, thus attracting some criticism of its human rights record.

ECONOMY
The Tunisian economy has shown significant improvement in recent years, recording steady growth. This prosperity has been led by the agricultural sector, which has benefited from good harvests in recent

years. The main crops are cereals and olives. The latter crop occupies one-third of arable land, and the nation is the world's fifth largest producer. In the south, using water from hot springs, tomatoes and melons are grown. This produce is exported in winter months when European Union (EU) quotas do not apply. Most trade is with EU countries and in early 1995 a partnership agreement was reached with the EU

Oil and gas also provide valuable revenue, partly through royalties from the trans-Mediterranean pipeline which delivers Algerian gas to Italy, and partly from its own resources. The development of the offshore Miskra gas field will further enhance this sector

The government has pursued a policy of liberalizing the economy. State assets are being privatized, but at a pace that is well-managed. The Tunisian government must avoid too rapid a change, that might exacerbate the uncomfortably high level of unemployment.

THE REPUBLIC OF TUNISIA	
STATUS	Republic
AREA	164,150 sq km (63,360 sq miles)
POPULATION	8,570,000
CAPITAL	Tunis
LANGUAGE	Arabic, French
RELIGION	Muslim
CURRENCY	Tunisian dinar (TND)
ORGANIZATIONS	Arab League, OAU, UN

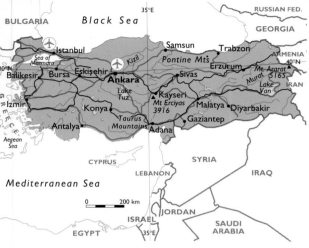

PHYSICAL GEOGRAPHY

Turkey, occupying the peninsula of Asia Minor, is dominated by two major systems of mountain ranges, the Pontine in the north and the Taurus in the south. Between the two is the high plateau of Anatolia.

CLIMATE

The coast has a Mediterranean climate, with mild winters and hot summers. The interior experiences great extremes, with hot dry summers, cold snowy winters and an average rainfall of less than 250 mm (10 inches) a year.

POLITICS AND RECENT HISTORY

After long periods of military domination, Turkey returned to civilian rule in 1983. In April 1993, President Turgut Özal died and Suleyman Demirel, formerly prime minister, was elected to succeed him. Tansu Çiller, leader of the True Path Party (DYP), became Turkey's first woman prime minister in June 1993, now leading a coalition government formed after the 1991 elections.

Turkey, at a three-way junction of Europe, Asia and the Middle East, is inevitably

involved in conflicts in those regions. Notably, it allowed its bases to be used by US forces in the Gulf War against Iraq. The southeast region is stricken with a worsening civil war between the Turkish army and Kurdish separatists of the Kurdistan Workers Party. Turkey applied to join the European Union (EU) in 1987, but for some years Greece stalled progress. However, a breakthrough was achieved in 1995 when Greece removed its veto and a customs union agreement was signed. The EU will, however, demand political and economic reform and an improvement in Turkey's human rights record before agreement to entry.

THE REPUBLIC OF TURKEY	
STATUS	Republic
AREA	779,450 sq km (300,870 sq miles)
POPULATION	58,775,000
CAPITAL	Ankara
LANGUAGE	Turkish, Kurdish
RELIGION	98% Sunni Muslim, Christian minority
CURRENCY	Turkish lira (TRL)
ORGANIZATIONS	Council of Europe, NATO, OECD, UN

ECONOMY

Textiles account for over a third of total exports, and the car industry is also important. In the agricultural sector tobacco, cotton, olives, citrus fruit, grapes and wheat are grown. Minerals exploited include copper, borax and chromium, although concern over the war against the Kurds has hindered expansion. Tourism has surged in importance as a major foreign currency earner since 1985.

It has been necessary for Tansu Çiller's government to take tough economic decisions in order to halt decline, and to start to meet conditions for EU entry. Although these have damaged her popularity and brought some hardship to the Turkish people, they are beginning to take effect. The economy is growing, inflation is falling fast and the trade deficit is much reduced.

While Turkey looks predominantly westwards towards the EU, whose countries take half its exports, it is also establishing links to the east. The Turkic speaking republics of the former Soviet Union have a natural affinity which has fostered close economic ties.

PHYSICAL GEOGRAPHY

Most of Turkmenistan is a lowland desert known as the Kara Kum; the only areas not thus classified are the shores of the Caspian Sea, the mountains along the border with Iran and the valley of the Amudar'ya along the northern boundary.

CLIMATE

Summers are hot, with maximum temperatures exceeding 30°C (86°F), while winter temperatures approach freezing point. Rainfall is sparse everywhere.

POLITICS AND RECENT HISTORY

Turkmenistan declared its independence from the Soviet Union in September 1991. This was recognized internationally when the Soviet Union dissolved in December 1991 and Turkmenistan became a member of the Commonwealth of Independent States. In the presidential elections of June 1992, Saparmurad Niyazov was elected, albeit no opposition candidates were permitted. Eighteen months later, a referendum authorized him to extend his presidency to 2002.

Elections are held to a People's Council. In December

1994, only one candidate was permitted in each of the 50 constituencies. Most of these were government nominees. The president wields strong executive power – his international political strategy is to strengthen Turkmenistan's links with nations to the south, such as Iran, Turkey and Pakistan, thus reducing ties with Russia.

ECONOMY

The republic enjoys economic wealth that should allow it to prosper. The basis of that wealth is very substantial oil and natural gas reserves. In addition, Turkmenistan has reserves of potassium, sulphur and salt.

The existence of the Kara Kum canal through its southern territories has allowed agriculture to thrive on the basis of irrigation. Cotton is the principal crop but cereals, fruit and melons are also grown.

Although the government retains strong centralist economic control it has, for example, designated seven free economic zones with exemptions from land tax and power supplies at reduced rates, in order to encourage foreign investment. The nation's stability is such that investment is likely to be forthcoming. The government has also initiated programme of privatization. The oil and gas industries are excluded and will remain under state control.

In August 1994, the republic signed an agreement with Iran and Turkey. Under its terms, natural gas will be exported via a pipeline through Iran. There are also plans for a rail link to Mashad in Iran. Further afield, Turkmenistan has been establishing trading links with European nations and has gained 'most favoured nation' status with the USA.

THE REPUBLIC OF TURKMENISTAN	
STATUS	Republic
AREA	488,100 sq km (188,405 sq miles)
POPULATION	3,921,000
CAPITAL	Ashkhabad (Ashgabat)
LANGUAGE	Turkmen, Russian, Uzbek
RELIGION	Muslim
CURRENCY	manat
ORGANIZATIONS	CIS, UN

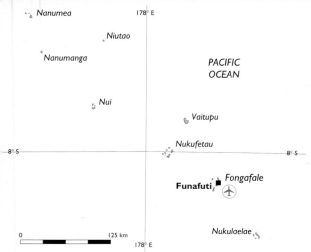

Nanumea

178° E

Niutao

Nanumanga

PACIFIC OCEAN

Nui

Vaitupu

8° S — *Nukufetau* — 8° S

Funafuti *Fongafale* ✈

Nukulaelae

0 — 125 km

178° E

PHYSICAL GEOGRAPHY

Tuvalu, one of the world's smallest nations, consists of nine coral atolls, all less than 4.6 m (15 ft) above sea-level. The total area is only 26 sq km (10 sq miles), and the islands are scattered over a vast region in the southern Pacific. One-third of the population lives on the island of Funafuti, on which is the capital and administrative centre.

CLIMATE

The climate is hot all the year round. Temperatures average 30°C (86°F), although March to October is a little cooler. The rainfall is heavy, averaging 3,050–4,050 mm (120–160 inches) annually, and falling mainly between November and February.

POLITICS AND RECENT HISTORY

The islands were once known as the Ellice Islands of the UK colony of the Gilbert and Ellice Islands partnership which, following a referendum, split in 1975. The Ellice Islands achieved independence in 1978 as Tuvalu. The Gilbert Islands became the Republic of Kiribati in 1979. Tuvalu is a

Niulakita

constitutional monarchy within the Commonwealth and the British monarch is head of state, represented locally by a governor-general. In 1986 a referendum was held on whether to change to a republican constitution, but the majority vote favoured retaining the monarchy. The government consists of a prime minister and four ministers selected from a 12-member assembly. In the most recent general election in November 1993, Kamuta Latasi became prime minister.

Tuvalu is one of the low-lying countries in the South Pacific which would be particu-

larly vulnerable should there be a rise in sea-level caused by the greenhouse effect. Following a United Nations report and discussions in the South Pacific Forum, it was agreed that a series of monitoring stations would be established in the region to gauge any significant climatic changes. Some believe that water levels are already rising. As soil is ruined by salt, farmers are moving inland into areas which are already overpopulated.

ECONOMY

The traditional occupations of the islanders are fishing and subsistence farming, but agriculture is restricted by poor soils and the only significant cash crop is coconuts for copra export. Other sources of income are from the sale of postage stamps and handicrafts, and remittances from Tuvaluans working abroad. Revenue is also raised from fishing licences issued to Japanese and American fleets operating in Tuvaluan waters. Tuvalu receives foreign financial assistance, including a grant from the UK towards road development, and the country is now a member of the Asian Development Bank.

TUVALU	
STATUS	State
AREA	24.6 sq km (9.5 sq miles)
POPULATION	9,000
CAPITAL	Funafuti
LANGUAGE	Tuvaluan, English
RELIGION	98% Protestant
CURRENCY	Australian dollar (AUD), Tuvaluan coinage
ORGANIZATIONS	Comm. (special member)

PHYSICAL GEOGRAPHY

Uganda is in the main a plateau with savannah vegetation, bordered in the west by the Ruwenzori Range and the Great Rift Valley. Southeast Uganda includes Lake Victoria, from which the Nile flows northwards through Lake Kyoga.

CLIMATE

The climate is tropical. Temperatures, though warm throughout the year, are modified by altitude. Rainfall occurs in all months.

POLITICS AND RECENT HISTORY

Uganda gained independence from Britain and joined the Commonwealth in 1962. In 1967 it became a republic. The president, Yoweri Museveni, came to power in a coup in 1986 and has governed the country since then through the National Resistance Movement. Uganda is in practice a one-party state, although Museveni describes it as a no-party state. Even so, the first signs of democracy emerged when, in early 1994, elections were held for a Constituent Assembly. This Assembly is to draft a Constitution and pave the way for parliamentary and presiden-

tial elections. The population showed their support for the president in these elections. He is trusted and respected, particularly in comparison with his corrupt predecessors Milton Obote and Idi Amin. He allows grass roots opinion to be voiced through a national network of resistance councils which have responsibility for local administration.

In July 1993, the president enhanced his popularity by allowing a tribal monarchy to be reinstated; Ronald Metebi was crowned King of the Baganda people 26 years after the monarchy was abolished. The king wields no power but is a powerful cultural symbol.

ECONOMY

President Museveni has brought stability and economic growth to Uganda, although, at the outset, under International Monetary Fund pressure he sacrificed 30,000 civil service jobs. More recently, the World Bank has provided funds to allow Uganda to pay off soldiers and reduce its armed forces.

The economy is now growing steadily and the Ugandan shilling is stable. The president appears to be well trusted internationally in that $800 million of foreign aid is forthcoming, despite the lack of democracy. Stability is also encouraging foreign investment. The decision to allow Ugandan Asians to return and reclaim their businesses has stimulated the economy and allowed entrepreneurial skills to flourish.

Nevertheless, it is agriculture that dominates the economy. Coffee is still the dominant crop and provides the greater part of Ugandan export revenue. There is, however, some diversification into tea, fruits and vegetables for the European market.

THE REPUBLIC OF UGANDA	
STATUS	Republic
AREA	236,580 sq km (91,320 sq miles)
POPULATION	19,940,000
CAPITAL	Kampala
LANGUAGE	English, tribal languages
RELIGION	62% Christian, 6% Muslim
CURRENCY	Uganda shilling (UGS)
ORGANIZATIONS	Comm., OAU, UN

PHYSICAL GEOGRAPHY
Much of the country is steppe, relatively flat and often forested.

CLIMATE
Ukraine has warm summers and cold winters, with milder conditions in the Crimea.

POLITICS AND RECENT HISTORY
The Ukrainian parliament declared independence from the Soviet Union in August 1991. This was confirmed by a referendum in December 1991, when the republic became a founder member of the Commonwealth of Independent States, and Leonid Kravchuk was elected president. His presidency ended at elections in July 1994, when Leonid Kuchma, who had once been Kravchuk's prime minister, won.

Following independence, relations with Russia were poor, but have gradually improved. The two governments have reached agreement in principle over the division of the Black Sea Fleet. Russian and western concerns over nuclear weapons have been allayed. The Ukrainian parliament has ratified the Strategic Arms Reduction Treaty and, more recently, endorsed the Nuclear Non-proliferation Treaty.

There is a wide gulf between the political aspirations of Ukrainian nationalists in the west and the pro-Russian Communists in the east and Crimea. In early 1995, President Kuchma dismissed the president of the semi-autonomous Crimea and imposed rule from Kiev.

ECONOMY
Ukraine was among the most successful of the Soviet Republics. In 1990 it provided 22 per cent of the USSR's agri-

UKRAINE	
STATUS	**Republic**
AREA	**603,700 sq km** **(233,030 sq miles)**
POPULATION	**52,000,000**
CAPITAL	**Kiev**
LANGUAGE	**Ukrainian, Russian**
RELIGION	**Russian Orthodox, Roman Catholic (Uniate)**
CURRENCY	**karbovanets (coupon)**
ORGANIZATIONS	**CIS, Council of Europe, UN**

CRIMEA	
STATUS	**Autonomous Republic of Ukraine**
AREA	**25,881 sq km** **(9,993 sq miles)**
POPULATION	**2,550,000**
CAPITAL	**Sevastopol**

cultural output and 16 per cent of industrial production. In the first few years of independence, the Ukraine came close to economic ruin as the government persisted in investing its meagre resources in the inefficient state-run businesses.

When President Kuchma gained office he immediately announced a programme of economic reform involving large-scale privatization, especially of land, cuts in subsidies and a general move towards a market economy. This change of direction has brought substantial financial support from the World Bank, International Monetary Fund and the European Union. As a result there is real hope that the Ukrainian economy will recover.

While the Ukrainian economy is inevitably closely linked to that of Russia, the government seeks to lessen this dependency. It recently announced the construction of an oil terminal at Odessa, allowing importation of oil from the Middle East. However, because of the shortage of other energy resources, Ukraine persists with nuclear energy, including the operation of the Chernobyl power station.

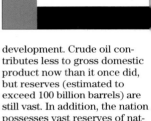

PHYSICAL GEOGRAPHY
Most of the United Arab Emirates is flat desert with sand dunes and salt pans. The only hilly area is in the northeast adjacent to the Gulf of Oman.

CLIMATE
Summers are hot and winters are mild. Coastal areas may receive a little rainfall in winter.

POLITICS AND RECENT HISTORY
The United Arab Emirates came into existence in 1971 when the United Kingdom relinquished responsibility for defence and foreign affairs. Initially six emirates came together, with Ras al Khaimah joining in 1972. The state is a federation in which the Supreme Council of Rulers is the main organ of government. This council comprises the hereditary rulers of each of the emirates, one of whom is elected by the others as President. Sheikh Zayed bin Sultan al Nahyan, the ruler of Abu Dhabi, has held this position since the foundation of the state. The Supreme Council appoints a 40-member advisory Council, but there are no political parties or parliamentary democracy.

The UAE government sets self-protection as a high priority and has invested large sums in defence. Its population is overwhelmingly made up of immigrants, and encouragement is offered to indigenous families to have children. Gradually, the state is taking a more international role and in 1993 contributed soldiers to the United Nations operations in Somalia.

ECONOMY
The economic wealth of the United Arab Emirates is founded on its reserves of hydrocarbons, mainly within the largest emirate, Abu Dhabi, with smaller supplies in three others – Dubai, Sharjah and Ras al Khaimah. Oil was first exported in 1962, and the revenue earned since then has allowed massive development. Crude oil contributes less to gross domestic product now than it once did, but reserves (estimated to exceed 100 billion barrels) are still vast. In addition, the nation possesses vast reserves of natural gas, approximately 5 per cent of the world's total, and is investing in liquefied natural gas facilities. The major market for its hydrocarbons is the Far East, Japan in particular.

Proper economic co-ordination between the seven constituent emirates has not always occurred, perhaps because of a degree of mutual suspicion. The result has been the duplication of facilities such as airports and hotels. Nevertheless, economic diversification has proceeded fast. Dubai has a thriving entrepôt trade, and desalination and irrigation have brought many acres into cultivation, allowing self-sufficiency in some vegetables and the export of strawberries and flowers to Europe.

Recent times have seen a drop in the Emirates' prosperity. The value of imports has risen sharply while oil exports have declined.

THE UNITED ARAB EMIRATES	
STATUS	Federation of seven Emirates
AREA	75,150 sq km (29,010 sq miles)
POPULATION	1,206,000
CAPITAL	Abu Dhabi (Abū Ẓabī)
LANGUAGE	Arabic, English
RELIGION	Sunni Muslim
CURRENCY	UAE dirham (AED)
ORGANIZATIONS	Arab League, OPEC, UN

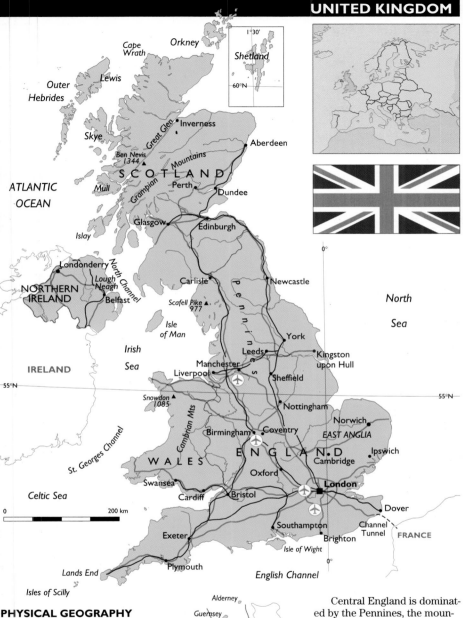

PHYSICAL GEOGRAPHY

The Highland zone of Britain consists of ancient uplifted rocks which now form the mountainous dissected and glaciated areas of the Lake District in the northwest and Wales, the Southern Uplands and Grampians of Scotland which rise to the highest point in the UK – 1,344 m (4,408 ft) at Ben Nevis. The latter are divided by the wide Central Lowland rift valley.

Central England is dominated by the Pennines, the mountain chain which stretches southwards from the Southern Uplands down the centre of England to the river Trent. The landscape of the southwest consists of the ancient uplifted granite domes of Dartmoor and Bodmin Moor.

Lowland Britain is a very contrasting landscape. Limestone and sandstone hills are separated by flat clay vales, east of a line joining the rivers Humber and Exe. Both the richest agricultural land and the densest population are found here.

Northern Ireland is generally hilly but has the UK's greatest lake, Lough Neagh, at its centre.

CLIMATE

The climate of the UK is mild, wet and variable. Summer temperatures average 13–17°C (55–63°F) and winter temperatures 5–7°C (41–45°F). Annual rainfall varies between 650 and 5,000 mm (26 and 200 inches) with the highest in the central Lake District and the lowest on the coasts of East Anglia.

POLITICS AND RECENT HISTORY

The United Kingdom, comprising England, Scotland, the principality of Wales and the province of Northern Ireland, is a constitutional monarchy with a parliamentary democracy, but there is no written constitution. Queen Elizabeth II, though head of state, exercises no real power. Although in recent years the monarchy has been the subject of some controversy, largely associated with family marriage breakdowns, it will in all probability survive.

At the last general election, held in 1992, the Conservative Party, under Prime Minister John Major, held on to power for a fourth term but with an absolute majority reduced to 21 seats over the Labour, Liberal Democrat and minor parties. This majority has since been further reduced as a result of losses in by-elections and the fall in popularity of the Conservative government was clearly demonstrated by their worst ever showing in local elections in May 1995. In contrast, the opposition Labour Party has been enjoying better fortunes. Tony Blair became leader, following the sudden death of John Smith in 1994. His most notable achievement within the party has been to win support for abandoning clause 4 of the Labour Party Constitution. This committed the party to seek state ownership of assets.

The most controversial issues on the British political agenda are those associated with membership of the European Union (EU). The government was able to win a narrow majority for ratification of the Maastricht Treaty but conflict continues, particularly within the Conservative Party, between those who favour closer links with European partners and those so-called Euro Sceptics who abhor any moves such as monetary union which, in their view, impinge on the nation's sovereignty.

Environmental issues are always important on the British political scene and are the cause of a greater emphasis on public transportation systems rather than the roads. Animal welfare issues, notably those concerning the export of live animals, have also fuelled controversy.

There can be no doubt, however, that the most significant political events of recent times have been those affecting the province of Northern Ireland. In December 1993 Major and the Irish Taoiseach issued the joint Downing Street Declaration. Although there was no immediate effect, by the autumn of 1994 Sinn Fein and the IRA had announced a cease-fire and an intention to participate in the political process. Progress after that was cautious, but in early 1995 the British and Irish governments signed a joint framework document as a basis for further discussions. This document caused anxiety among the Unionist sections of the community, and it did not meet the aspirations of Nationalists. May 1995 saw the first ever official meeting between Sinn Fein and a British government minister. However by early 1996 the IRA had resumed its bombing campaign.

ECONOMY

Although a small percentage of the nation's workforce are employed in agriculture, farm produce is important to both home and export markets. Seventy-six per cent of the total UK land area is farmland. The main cereal crops are wheat, barley and oats. Potatoes, sugar-beet and green vegetable crops are widespread.

About 20 per cent of the land is permanent pasture for the raising of dairy and beef stock, and 28 per cent of the land, mainly hill and mountain areas, is used for rough grazing of sheep. Pigs and poultry are widespread in both England and lowland Scotland. The best fruit-growing areas are the southeast, especially Kent, and East Anglia and the central Vale of Evesham for apples, pears and soft fruit. Both the forestry and fishing industries contribute to the economy.

The major mineral resources of the UK are coal, oil and natural gas. Coal output goes towards the generation of

THE UNITED KINGDOM OF GREAT BRITAIN AND NORTHERN IRELAND	
STATUS	Kingdom
AREA	244,755 sq km (94,475 sq miles)
POPULATION	57,998,400
CAPITAL	London
LANGUAGE	English, Welsh, Gaelic
RELIGION	Protestant majority. Roman Catholic, Jewish, Muslim, Hindu minorities
CURRENCY	pound sterling (GBP)
ORGANIZATIONS	Col. Plan, Comm., Council of Europe, EEA, EU, G7, NATO, OECD, UN, WEU

electricity but oil and natural gas from the North Sea, and to a lesser extent nuclear power, are divided between the needs of industry and the consumer. Iron ore, once mined sufficiently to satisfy industry, is now imported to support the iron and steel manufacturing sector.

The UK produces a great range of industrial goods for home consumption and export. General and consumer goods manufacturing is located in all heavy industrial areas but the London area, the West Midlands, Lancashire and Merseyside predominate. Main products are food and drinks, chemicals, light engineering products, cotton and woollen textiles, electrical and electronic goods.

The UK is a trading nation. The balance of trade has changed during the last 30 years because of stronger economic, military and political ties within Europe – the EU and the North Atlantic Treaty Organization – and consequently reduced trading links with former colonies, particularly in Australasia. Major exports are cereals, meat, dairy products, beverages, tobacco products, textiles, metalliferous ores, petroleum and petroleum products, chemicals, pharmaceutical goods, plastics, leather goods, rubber, paper, iron and steel, other metal goods, engines and vehicles, machinery, electrical and electronic goods and transport equipment.

The UK has a highly developed transport network. Motorways, trunk roads and principal roads total over 50,000 km (31,070 miles). The railway network covers 16,730 km (10,395 miles) and now carries over 140 million tonnes of freight annually. The inland waterway system, once a major freight carrier, totals only 563 navigable kilometres (350 miles) but still carries over four million tonnes of goods annually.

The UK was unable to stay within the European exchange rate mechanism, but since its withdrawal in 1992 the British economy has shown a gradual recovery, although some sectors of the economy such as the construction industry remain depressed. Indeed, in 1994 the economy grew faster than had been predicted and prompted concern about inflationary pressures. Arguably the most serious economic problem remains the high rate of unemployment. The figure has shown a steady reduction but still imposes a heavy burden on public funds.

One of the more encouraging aspects of the UK's economy is that it continues to attract a major share of inward investment, especially from the USA and Japan. In total a third of all investment into the EU has been to the United Kingdom.

Britain's oil industry has been a major source of revenue since the 1970s. Although exploitation of North Sea reserves will be important for many years to come, emphasis is beginning to shift towards exploration in other areas such as the seas to the west of the Shetlands.

The biggest single infrastructure development was brought to fruition in 1995, when the Channel Tunnel was fully opened for both freight and passenger services. Full economic benefits will, however, only accrue when high speed rail links are developed at least between the Tunnel and London, and a possible second link to the Midlands.

ENGLAND	
STATUS	Constituent Country
AREA	130,360 sq km (50,320 sq miles)
POPULATION	48,378,300
CAPITAL	London

NORTHERN IRELAND	
STATUS	Constituent Region
AREA	14,150 sq km (5,460 sq miles)
POPULATION	1,589,000
CAPITAL	Belfast

SCOTLAND	
STATUS	Constituent Country
AREA	78,750 sq km (30,400 sq miles)
POPULATION	4,957,000
CAPITAL	Edinburgh

WALES	
STATUS	Principality
AREA	20,760 sq km (8,015 sq miles)
POPULATION	2,898,500
CAPITAL	Cardiff

ISLE OF MAN

STATUS .....**British Crown Dependency**

AREA ....**588 sq km**
(227 sq miles)

POPULATION.....................**71,000**

CAPITAL.........................**Douglas**

GUERNSEY

STATUS ..**Island of the Channel Islands**

AREA...........................**65 sq km**
(25 sq miles)

POPULATION.....................**58,867**

CAPITAL....................**St Peter Port**

JERSEY

STATUS ..**Island of the Channel Islands**

AREA ....**116 sq km**
(45 sq miles)

POPULATION.....................**84,082**

CAPITAL.........................**St Helier**

GIBRALTAR

STATUS ....**UK Crown Colony**

AREA ....**6.5 sq km**
(2.5 sq miles)

POPULATION.....................**31,000**

CAPITAL ....**Gibraltar**

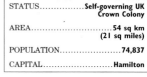

FALKLAND ISLANDS

STATUS ....**UK Dependent Territory**

AREA......................**12,175 sq km**
(4,700 sq miles)

POPULATION **2,121**

CAPITAL ....**Stanley**

MONTSERRAT

STATUS ....**UK Dependent Territory**

AREA ....**106 sq km**
(41 sq miles)

POPULATION.....................**13,000**

CAPITAL ....**Plymouth**

BERMUDA

STATUS................**Self-governing UK**
Crown Colony

AREA...........................**54 sq km**
(21 sq miles)

POPULATION.....................**74,837**

CAPITAL........................**Hamilton**

ANGUILLA

STATUS ....**UK Dependent Territory**

AREA ....**155 sq km**
(60 sq miles)

POPULATION **8,960**

CAPITAL.....................**The Valley**

Turks and Caicos Islands
British Virgin Islands
Cayman Islands
Anguilla
Hong Kong
St Helena & Dependencies
British Indian Ocean Territory
Pitcairn Island
South Georgia

BRITISH VIRGIN ISLANDS

STATUS ....**UK Dependent Territory**

AREA ....**153 sq km**
(59 sq miles)

POPULATION.....................**16,749**

CAPITAL.....................**Road Town**

CAYMAN ISLANDS

STATUS ....**UK Dependent Territory**

AREA ....**259 sq km**
(100 sq miles)

POPULATION.....................**29,000**

CAPITAL ....**George Town**

HONG KONG

STATUS ....**UK Dependent Territory**
(includes Kowloon and the New Territories)

AREA ....**1,067 sq km**
(412 sq miles)

POPULATION.................**5,919,000**

CAPITAL..........................**Victoria**

PITCAIRN ISLAND

STATUS ....**UK Dependent Territory**

AREA...........................**45 sq km**
(17.25 sq miles)

POPULATION...........................**71**

CAPITAL.....................**Adamstown**

ST HELENA

STATUS ....**UK Dependent Territory**

AREA ....**122 sq km**
(47 sq miles)

POPULATION **5,700**

CAPITAL......................**Jamestown**

TURKS AND CAICOS ISLANDS

STATUS ....**UK Dependent Territory**

AREA ....**430 sq km**
(166 sq miles)

POPULATION.....................**11,696**

CAPITAL ....**Cockburn Town**

BRITISH INDIAN OCEAN TERRITORY

STATUS..................**UK Dependency**
comprising the Chagos Archipelago

AREA...........................**52 sq km**
(20 sq miles)

POPULATION **2,900**

CAPITAL ....**none**

SOUTH GEORGIA AND THE SOUTH SANDWICH ISLANDS

STATUS ....**UK Dependent Territory**

AREA ....**3,755 sq km**
(1,450 sq miles)

POPULATION **1,450**

CAPITAL ....**none**

and cold winters. Only along the Pacific coast is the maritime influence strong.

PHYSICAL GEOGRAPHY

The eastern states, originally forested and still largely tree covered, include the Atlantic coastal plain, which widens southwards, and the Appalachian mountains running from the northeast extremity to Georgia. The central states form the Great Plains, with prairie vegetation drained by the Mississippi–Missouri river system. To the west lie the Rocky Mountains, separated from the Pacific coastal ranges by the great intermontane basin. The coastal ranges extend northwards and are a feature of the state of Alaska. The 50th state, Hawaii, is an archipelago of some 20 volcanic islands.

CLIMATE

Climatic conditions are tropical in the south and southeast, often hot and dry in the southwest and more temperate elsewhere. Much of the country experiences marked seasonal variation, with warm summers

POLITICS AND RECENT HISTORY

The written constitution of the United States provides for a federal form of government in which there is a separation of executive and legislative power between the president and a two-chamber congress. In the presidential elections of November 1992 Bill Clinton, the Democratic candidate, defeated Republican incumbent George Bush and was installed in January 1993. President Clinton suffered a major setback at the mid-term

THE UNITED STATES OF AMERICA	
STATUS	Federal Republic
AREA	9,363,130 sq km (3,614,170 sq miles)
POPULATION	255,020,000
CAPITAL	Washington D.C.
LANGUAGE	English, Spanish
RELIGION	Christian majority, Jewish minority
CURRENCY	US dollar (USD)
ORGANIZATIONS	ANZUS, Col. Plan, G7, NAFTA, NATO, OAS, OECD, UN

Congressional elections in November 1994 when Republicans, who had been in a minority in both Senate and House of Representatives, gained a majority in both for the first time in 40 years. Newt Gingrich became Speaker of the House and Senator Bob Dole leader in the Senate.

The first two years of the Clinton administration had in any case not been easy. He had been forced to concede on a number of issues, such as admission of homosexuals to the armed forces and most importantly, his flagship proposals on health care. Clinton's main successes were approval for the North American Free Trade Agreement (NAFTA) and his budget bill.

However, it is still the budget that occupies Congress the most and the need to correct the long-standing deficit. President Clinton introduced his own proposals which aim to achieve a balance over ten years and include some spending cuts on health care and modest tax cuts. However, these gave rise to serious conflict with the Republican-dominated Congress. In foreign policy the involvement of American troops in Somalia proved disastrous. Strenuous efforts to restore President Aristide in Haiti were however successful. After steadfastly refusing to commit US troops to the UN operations in Bosnia, the US government brokered the important Dayton peace agreement in December 1995.

The United States continues to be plagued by illegal immigrants from Mexico. In the first quarter of 1995, some 360,000 arrests were made along the border. The government has also clamped down on immigration from Cuba.

The next presidential elections are due in 1996. Senator Dole is certain to provide President Clinton with a strong challenge.

ECONOMY

The USA consumes 25 per cent of all the world's energy resources but is well endowed with energy reserves. There are substantial coal resources in Pennsylvania, the Appalachian region, the Dakotas and Wyoming, and oil and natural gas regions in Texas, Louisiana, Alaska, and the Gulf of Mexico. However, the nation now imports more than 50 per cent of its oil needs. This situation is viewed with some anxiety and is why tax incentives are offered to oil prospectors. The vast resources of the great rivers have been harnessed extensively for hydro-electric power. In the west, mineral deposits include copper, lead, zinc and silver, and there is iron ore around Lake Superior. Most specialist industrial minerals are imported. Diamonds, tin, chromite, nickel, asbestos, platinum, manganese, mercury, tungsten, cobalt, antimony and cadmium are not found in sufficient quantities for home

demand. Non-metallic minerals extracted within the USA include clays, gypsum, lime, phosphate, salt, sand, gravel and sulphur.

America's first industrialized area lies to the south of the Great Lakes and has gradually extended to form one of the largest industrial zones in the world. Chicago is the main steel-producing town, and Pennsylvania and Pittsburgh are famous for their steel and chemical industries. Along the west coast the main industries include vehicle manufacture, armaments and electrical and electronic goods.

The US economy is gradually recovering from recession. gross domestic product is showing modest growth at about 3 per cent, inflation is under control and unemployment, though only just below 7 per cent, is diminishing. The trade balance is, however, somewhat unhealthy with a continuing deficit. Despite the attentions of the president and congress, there is no credible solution in sight for the most deep rooted problem of the American economy, the budget deficit. The political imperatives of all those involved do not yet allow the necessary measures to be taken.

Congressional approval for NAFTA may be seen as an economic landmark as it promises to free up trade between the USA, Canada and Mexico. It offers opportunities for job creation in all three countries but was viewed with suspicion by protectionist elements. The bill's passage through congress was dependent more on Republican than Democrat support and was only possible because President Clinton made concessions to particularly interested groups such as wine producers, fruit and vegetable growers and wheat farmers near the Canadian border.

Whatever difficulties the US

government may face, the nation's economy is unrivalled in its power and strength. The American people enjoy, on average, the highest standard of living anywhere in the world.

ALABAMA	
STATUS	State
AREA	131,485 sq km (50,755 sq miles)
POPULATION	4,089,000
CAPITAL	Montgomery

ALASKA	
STATUS	State
AREA	1,478,450 sq km (570,680 sq miles)
POPULATION	570,000
CAPITAL	Juneau

ARIZONA	
STATUS	State
AREA	293,985 sq km (113,480 sq miles)
POPULATION	3,750,000
CAPITAL	Phoenix

ARKANSAS
STATUS...................................State

AREA.....................134,880 sq km
(52,065 sq miles)

POPULATION.................2,372,000

CAPITAL......................Little Rock

CALIFORNIA
STATUS...................................State

AREA.....................404,815 sq km
(156,260 sq miles)

POPULATION...............30,380,000

CAPITAL....................Sacramento

COLORADO
STATUS...................................State

AREA.....................268,310 sq km
(103,570 sq miles)

POPULATION.................3,377,000

CAPITAL...........................Denver

CONNECTICUT
STATUS...................................State

AREA........................12,620 sq km
(4,870 sq miles)

POPULATION.................3,291,000

CAPITAL..........................Hartford

DELAWARE
STATUS...................................State

AREA.........................5,005 sq km
(1,930 sq miles)

POPULATION....................680,000

CAPITAL..............................Dover

DISTRICT OF COLUMBIA
STATUS..................Federal District

AREA............................163 sq km
(63 sq miles)

POPULATION....................598,000

CAPITAL.....................Washington

FLORIDA
STATUS...................................State

AREA.....................140,255 sq km
(54,140 sq miles)

POPULATION...............13,277,000

CAPITAL.....................Tallahassee

GEORGIA
STATUS...................................State

AREA.....................150,365 sq km
(58,040 sq miles)

POPULATION.................6,623,000

CAPITAL............................Atlanta

HAWAII
STATUS...................................State

AREA........................16,640 sq km
(6,425 sq miles)

POPULATION.................1,135,000

CAPITAL..........................Honolulu

IDAHO
STATUS...................................State

AREA.....................213,445 sq km
(82,390 sq miles)

POPULATION.................1,039,000

CAPITAL..............................Boise

ILLINOIS
STATUS...................................State

AREA.....................144,120 sq km
(55,630 sq miles)

POPULATION...............11,543,000

CAPITAL.....................Springfield

INDIANA
STATUS...................................State

AREA........................93,065 sq km
(35,925 sq miles)

POPULATION.................5,610,000

CAPITAL....................Indianapolis

IOWA
STATUS...................................State

AREA.....................144,950 sq km
(55,950 sq miles)

POPULATION.................2,795,000

CAPITAL.....................Des Moines

KANSAS
STATUS...................................State

AREA.....................211,805 sq km
(81,755 sq miles)

POPULATION.................2,495,000

CAPITAL.............................Topeka

KENTUCKY
STATUS...................................State

AREA.....................102,740 sq km
(39,660 sq miles)

POPULATION.................3,713,000

CAPITAL..........................Frankfort

LOUISIANA
STATUS...................................State

AREA.....................115,310 sq km
(44,510 sq miles)

POPULATION.................4,252,000

CAPITAL.....................Baton Rouge

MAINE
STATUS...................................State

AREA........................80,275 sq km
(30,985 sq miles)

POPULATION.................1,235,000

CAPITAL............................Augusta

MARYLAND
STATUS...................................State

AREA........................25,480 sq km
(9,835 sq miles)

POPULATION.................4,860,000

CAPITAL..........................Annapolis

MASSACHUSETTS
STATUS...................................State

AREA........................20,265 sq km
(7,820 sq miles)

POPULATION.................5,996,000

CAPITAL.............................Boston

MICHIGAN
STATUS...................................State

AREA.....................147,510 sq km
(56,940 sq miles)

POPULATION.................9,368,000

CAPITAL...........................Lansing

MINNESOTA
STATUS...................................State

AREA.....................206,030 sq km
(79,530 sq miles)

POPULATION.................4,432,000

CAPITAL............................St Paul

MISSISSIPPI
STATUS...................................State

AREA.....................122,335 sq km
(47,220 sq miles)

POPULATION.................2,592,000

CAPITAL............................Jackson

MISSOURI
STATUS...................................State

AREA.....................178,565 sq km
(68,925 sq miles)

POPULATION.................5,158,000

CAPITAL....................Jefferson City

MONTANA
STATUS...................................State

AREA.....................376,555 sq km
(145,350 sq miles)

POPULATION....................808,000

CAPITAL.............................Helena

NEBRASKA
STATUS .... State

AREA .... 198,505 sq km
(76,625 sq miles)

POPULATION 1,593,000

CAPITAL .... Lincoln

NEVADA
STATUS .... State

AREA .... 284,625 sq km
(109,865 sq miles)

POPULATION 1,284,000

CAPITAL .... Carson City

NEW HAMPSHIRE
STATUS .... State

AREA .... 23,290 sq km
(8,990 sq miles)

POPULATION 1,105,000

CAPITAL .... Concord

NEW JERSEY
STATUS .... State

AREA .... 19,340 sq km
(7,465 sq miles)

POPULATION 7,760,000

CAPITAL .... Trenton

NEW MEXICO
STATUS .... State

AREA .... 314,255 sq km
(121,300 sq miles)

POPULATION 1,548,000

CAPITAL .... Sante Fe

NEW YORK
STATUS .... State

AREA .... 122,705 sq km
(47,365 sq miles)

POPULATION 18,058,000

CAPITAL .... Albany

NORTH CAROLINA
STATUS .... State

AREA .... 126,505 sq km
(48,830 sq miles)

POPULATION 6,737,000

CAPITAL .... Raleigh

NORTH DAKOTA
STATUS .... State

AREA .... 179,485 sq km
(69,280 sq miles)

POPULATION 635,000

CAPITAL .... Bismarck

OHIO
STATUS .... State

AREA .... 106,200 sq km
(40,995 sq miles)

POPULATION 10,939,000

CAPITAL .... Columbus

OKLAHOMA CITY
STATUS .... State

AREA .... 177,815 sq km
(68,635 sq miles)

POPULATION 3,175,000

CAPITAL .... Oklahoma City

OREGON
STATUS .... State

AREA .... 249,115 sq km
(96,160 sq miles)

POPULATION 2,922,000

CAPITAL .... Salem

PENNSYLVANIA
STATUS .... State

AREA .... 116,260 sq km
(44,875 sq miles)

POPULATION 11,961,000

CAPITAL .... Harrisburg

RHODE ISLAND
STATUS .... State

AREA .... 2,730 sq km
(1,055 sq miles)

POPULATION 1,004,000

CAPITAL .... Providence

SOUTH CAROLINA
STATUS .... State

AREA .... 78,225 sq km
(30,195 sq miles)

POPULATION 3,560,000

CAPITAL .... Columbia

SOUTH DAKOTA
STATUS .... State

AREA .... 196,715 sq km
(75,930 sq miles)

POPULATION 703,000

CAPITAL .... Pierre

TENNESSEE
STATUS .... State

AREA .... 106,590 sq km
(41,145 sq miles)

POPULATION 4,953,000

CAPITAL .... Nashville

TEXAS
STATUS .... State

AREA .... 678,620 sq km
(261,950 sq miles)

POPULATION 17,349,000

CAPITAL .... Austin

UTAH
STATUS .... State

AREA .... 212,570 sq km
(82,050 sq miles)

POPULATION 1,770,000

CAPITAL .... Salt Lake City

VERMONT
STATUS .... State

AREA .... 24,015 sq km
(9,270 sq miles)

POPULATION 567,000

CAPITAL .... Montpelier

VIRGINIA
STATUS .... State

AREA .... 102,835 sq km
(39,695 sq miles)

POPULATION 6,286,000

CAPITAL .... Richmond

WASHINGTON
STATUS .... State

AREA .... 172,265 sq km
(66,495 sq miles)

POPULATION 5,018,000

CAPITAL .... Olympia

WEST VIRGINIA
STATUS .... State

AREA .... 62,470 sq km
(24,115 sq miles)

POPULATION 1,801,400

CAPITAL .... Charleston

WISCONSIN
STATUS .... State

AREA .... 140,965 sq km
(54,415 sq miles)

POPULATION 4,955,000

CAPITAL .... Madison

WYOMING
STATUS .... State

AREA .... 251,200 sq km
(96,965 sq miles)

POPULATION 460,000

CAPITAL .... Wyoming

ALEUTIAN ISLANDS

STATUS.................**Territory of USA**

AREA........................**17,665 sq km**
(6,820 sq miles)

POPULATION......................**11,942**

CAPITAL...............................**none**

AMERICAN SAMOA

STATUS ......**Unincorporated Territory of USA**

AREA ....**197 sq km**
(76 sq miles)

POPULATION......................**46,773**

CAPITAL.........................**Pago Pago**

GUAM

STATUS ....**External Territory of USA**

AREA ....**450 sq km**
(174 sq miles)

POPULATION **132,726**

CAPITAL..............................**Agana**

PUERTO RICO

STATUS ..**Self-governing Commonwealth of USA**

AREA ....**8,960 sq km**
(3,460 sq miles)

POPULATION **3,580,000**

CAPITAL..........................**San Juan**

AMERICAN VIRGIN ISLANDS

STATUS........**External Territory of USA**

AREA ....**345 sq km**
(133 sq miles)

POPULATION **101,809**

CAPITAL...............**Charlotte Amalie**

NORTHERN MARIANA ISLANDS

STATUS .**Self-governing Commonwealth of USA**

AREA ....**471 sq km**
(182 sq miles)

POPULATION......................**43,345**

CAPITAL.............................**Saipan**

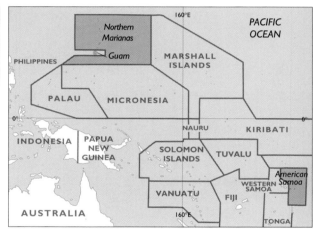

PHYSICAL GEOGRAPHY

Uruguay is a low-lying land of prairies, reaching only 500 m (1,600 ft) above sea-level at its highest point. The Atlantic coast and the estuary of the River Plate in the south are fringed with lagoons and sand dunes.

CLIMATE

The climate is one of equable warm summers and mild winters. Moderate rainfall is spread evenly at about 100 mm (4 inches) through each month of the year.

POLITICS AND RECENT HISTORY

Civilian government returned to Uruguay in 1984, following 12 years of military rule. Politics had been dominated by two political parties, the right wing *Partido Nacional*, known as the *Blancos*, and the centre-left Colorado party. Elections in late 1989 secured victory for the *Blancos*, and their leader Luis Alberto Lacalle assumed the presidency in the following March. However, the government had not secured an over-all majority and formed a pact known as the *Coincidencia Nacional* with some Colorado

factions. This uneasy relation-ship has not led to decisive rule. At elections in November 1994, a new party, the *Encuentro Progresista* emerged and the three parties fought a battle in which a virtu-al three-way tie resulted. The narrow victor was the Colorado candidate Julio Mario Sanguinetti.

President Sanguinetti's coalition government will find it difficult to introduce eco-nomic reform. The previous administration's attempts to introduce privatization were brought to a halt by a referen-dum. The population shows more concern about the pro-tection of the generous welfare system.

ECONOMY

The Uruguayan economy is

THE ORIENTAL REPUBLIC OF URUGUAY	
STATUS	Republic
AREA	186,925 sq km (72,155 sq miles)
POPULATION	3,149,000
CAPITAL	Montevideo
LANGUAGE	Spanish
RELIGION	Roman Catholic
CURRENCY	Uruguayan peso (UYP)
ORGANIZATIONS	Mercosur, OAS, UN

based overwhelmingly on agri-culture. Stock rearing predomi-nates and 85 per cent of the land area is devoted to this industry. Meat and wool are thus the most important exports and source of revenue. The milk industry has been successful and has been given a boost since the Italian com-pany, Parmalat, has invested in it. Their involvement will ensure greater competitiveness from the national co-operative Conaprole, which currently absorbs 80 per cent of output.

The Uruguayan economy has been showing steady growth, unemployment is falling and membership of Mercosur improves prospects as the common market devel-ops. However, the currency appears overvalued and has led to a substantial trade deficit. Any attempt to devalue the peso must take account of inflation which, although grad-ually decreasing, is around 40 per cent.

PHYSICAL GEOGRAPHY

Most of the country comprises the flat desert of the Kyzyl Kum, which rises eastwards towards the mountains of the western Pamirs. The Aral Sea in the north has shrunk by one third over the past 20 years, due to demand for irrigation water drawn from its feeding rivers.

CLIMATE

The climate is dry and arid, with high summer temperatures often reaching over 50°C (120°F).

POLITICS AND RECENT HISTORY

Although independent since 1991, Uzbekistan in some ways remains the least changed of the former Soviet republics. Islam Karimov, the ex-Communist party ruler, became President in December 1991, defeating just one other candidate, and the former Communists were renamed the People's Democratic Party (PDP). In March 1993 parliament approved a new constitution.

Local papers are censored and the leading Moscow newspaper, *Izvestia*, is banned. In December 1991 President Karimov refused to sign the parts of the Commonwealth of Independent States charter which guarantee respect for human rights.

A general election, nominally democratic, was held at the end of 1994. However, true opposition parties have been effectively eliminated by oppression and the PDP won an overwhelming majority. The few seats that did no go to the PDP went to officially authorized parties that are little different. A few months later, a compliant population voted in a referendum to allow President Karimov to extend his six-year term of office, due to expire in 1997, to the end of the century.

ECONOMY

Uzbekistan, with a basically agricultural economy, is the world's third largest cotton producer. It accounts for 67 per cent of the cotton grown in the region of the former USSR, but sales to the old Soviet republics are coming to a halt and the Uzbeks plan to market the cotton on a barter basis to western countries. In order to ensure greater food supplies, less land is being devoted to cotton and more to cereals. Tobacco is also an important crop which has attracted investment from the British American Tobacco company.

Uzbekistan is the eighth largest producer of gold. This together with the country's gas reserves, has attracted involvement from overseas.

The outlook for the Uzbekistan economy, with its combination of natural resources, agriculture and a relatively efficient fertilizer industry, is promising. In the short-term, help is necessary and negotiations with the International Monetary Fund have prompted some economic reform. Meanwhile, the population have yet to benefit.

THE REPUBLIC OF UZBEKISTAN	
STATUS	Republic
AREA	447,400 sq km (172,695 sq miles)
POPULATION	21,700,000
CAPITAL	Tashkent
LANGUAGE	Uzbek, Russian, Turkish
RELIGION	Muslim
CURRENCY	sum
ORGANIZATIONS	CIS, UN

KARAKALPAK (KARAKALPAKSTAN)	
STATUS	Autonomous Republic of Uzbekistan
AREA	164,900 sq km (63,920 sq miles)
POPULATION	169,000
CAPITAL	Nukus

PHYSICAL GEOGRAPHY

Vanuatu is an archipelago of some 80 islands, many of which are densely forested and mountainous. There are five active volcanoes.

CLIMATE

The climate is tropical with high rainfall and a continual threat of cyclones.

POLITICS AND RECENT HISTORY

Vanuatu became independent, from its former status as the Anglo-French Condominium of New Hebrides, in 1980. Politics are controlled by a coalition in which the francophone Union of Moderate Parties is dominant. Recent trends have seen a strengthening of relations with France, although major aid programmes have been agreed with Australia. The republic was hit by Cyclone Prema in March 1993, which destroyed plantations and whole villages.

ECONOMY

Copra accounts for 75 per cent of Vanuatu's export earnings, and cocoa and fish are also important products. Tourism, based on attractions such as the active volcano on Tanna, is expanding. The tourist trade, mostly Australian and Japanese, is assisted by the national airline, Air Vanuatu. There are plans to expand the airport at Port-Vila to take

THE REPUBLIC OF VANUATU	
STATUS	Republic
AREA	14,765 sq km (5,700 sq miles)
POPULATION	156,000
CAPITAL	Port-Vila
LANGUAGE	Bislama (national), English, French, Melanesian languages
RELIGION	Christian
CURRENCY	vatu (VUV)
ORGANIZATIONS	Comm., UN

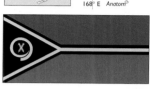

jumbo jets from Australia and Japan. French influence can be seen in the development of the telecommunications system, road improvement and the spread of the French language.

PHYSICAL GEOGRAPHY

Vatican City, occupying a hill to the west of the river Tiber, is located entirely within Rome, but is separated by a surrounding wall.

POLITICS AND RECENT HISTORY

The Vatican City State gained formal recognition from Italy as an independent base for the Holy See – the government of the Roman Catholic Church – in 1929. Sovereignty and executive power are exercised by the Pope, John Paul II, who was inaugurated in 1978, although administration is delegated to a commission.

On 30 December 1993 a 'Fundamental Agreement between the Holy See and the State of Israel' was signed in Jerusalem, establishing diplomatic relations between the two states. After 17 months of negotiations this brought a formal end to an historic enmity between the Church of Rome and the Jewish people. In 1994, the Vatican also established diplomatic relations with Jordan.

ECONOMY

The main sources of income in the Vatican City are the Institute of Religious Works, interest on the Vatican's investments and voluntary contributions (Peter's Pence). Tourism is important, the attractions including Michelangelo's works

THE VATICAN CITY	
STATUS	Ecclesiastical State
AREA	0.44 sq km (0.17 sq miles)
POPULATION	1,000
CAPITAL	none
LANGUAGE	Italian, Latin
RELIGION	Roman Catholic
CURRENCY	Italian lira (ITL), Papal coinage
ORGANIZATIONS	UN observer

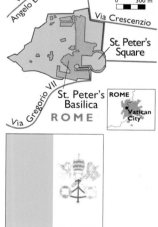

in the Sistine Chapel.

The Vatican's published accounts have always previously shown a deficit, but for the year to mid-1994, the accounts recorded a profit for the first time.

PHYSICAL GEOGRAPHY

Mountain ranges, initially running parallel to the Caribbean coast and then turning south-westwards, form the most northerly extension of the Andes chain and traverse the country. To the south and east of these mountains there are lowland grasslands drained by the Orinoco river system, and to the northwest are the lowlands around Lake Maracaibo.

CLIMATE

Temperatures are warm throughout the year, although the figures are dependent upon altitude. The wettest months are in summer.

POLITICS

Venezuela freed itself from military dictatorship in 1958, since when government has alternated between the Christian Democrat and Democratic Action parties. However, the president wields substantial authority. When Carlos Andrés Peréz was installed as president in 1989, he introduced a programme of economic reform. Such was the unpopularity that military coups were attempted and, though they failed, he was forced out of office and impeached for corruption.

Rafael Caldera led a coalition of small parties, the National Convergence coalition, to victory in the elections that followed at the end of 1993. He is popular with the electorate, having served as president from 1969 to 1974 under a different banner. He came to power with a promise to root out corruption but since the judiciary itself has been shown to be corrupt, he will face some difficulties. Since taking office he has been accused of anti-democratic measures in limiting the powers of parliament.

ECONOMY

The Venezuelan economy was founded on oil wealth, but reductions in world prices have led to harsher times. The economy has been in recession, there is an acute budgetary deficit and inflation is high.

When Peréz came to power, he introduced unpopular economic reforms that contributed to his downfall. President Caldera has now been forced to introduce similar measures, although his popularity may allow him to carry them through. He has introduced a two-year plan involving privatization and cuts in government subsidies, and has won qualified support from the International Monetary Fund, with the proviso that his short-term crisis measures must be transformed into a long-term strategy.

THE REPUBLIC OF VENEZUELA	
STATUS	Republic
AREA	912,045 sq km (352,050 sq miles)
POPULATION	20,712,000
CAPITAL	Caracas
LANGUAGE	Spanish
RELIGION	Roman Catholic
CURRENCY	bolivar (VEB)
ORGANIZATIONS	OAS, OPEC, UN

PHYSICAL GEOGRAPHY

The Red River delta lowlands are separated from the Mekong delta in the south by generally rough mountainous terrain.

CLIMATE

The climate is tropical, with temperatures varying from 30°C (86°F) in summer to less than 20°C (68°F) in winter. Summer monsoon rains fall from May to October; winters are relatively dry.

POLITICS AND RECENT HISTORY

Communist rule has persisted since the withdrawal of US troops from the south in 1975, and Vietnam remains a one-party state. Although the government initiated a programme of economic reform during the 1990s, there is little sign of this spreading to the political scene. The head of state is President General Le Duc Anh but real power is wielded by the Communist Party secretary Do Muoi.

The Vietnamese government cracks down hard on any internal dissent but has been seeking to establish good relations with its neighbours. In particular, there is a new found friendship with China, its traditional adversary. The two nations have agreed to set up a committee to resolve territorial disputes in the South China Sea. There are also moves towards a restoration of relations with the United States, who severed diplomatic relations in 1975. The US government has opened a diplomatic liaison office and it is expected that ambassadors will be exchanged before long.

ECONOMY

Following years of stagnation, the Vietnamese government recently embarked upon a programme of change. The econo-

my is now recovering and has exhibited consistent growth. Industrial output has shown dramatic improvement with steel production and car manufacturing leading. The oil and gas industry is booming, following the announcement of

THE SOCIALIST REPUBLIC OF VIETNAM	
STATUS	Republic
AREA	329,566 sq km (127,246 sq miles)
POPULATION	71,324,000
CAPITAL	Hanoi
LANGUAGE	Vietnamese, French, Chinese
RELIGION	Buddhist
CURRENCY	dong (VND)
ORGANIZATIONS	ASEAN, UN

recent offshore finds by both BP and the Mitsubishi Corporation.

Rice production has expanded such that Vietnam, once an importer, is now the world's third leading exporter after the USA and Thailand.

Despite this, Vietnam is a poor country. The United Nations estimate that 51 per cent of the population live below the poverty level. The situation is worse in the countryside where economic reforms are slow to bring benefit. The effect has been a drift to the towns, although the population is still predominantly rural.

However, economic development is, however, such that large-scale poverty should soon be eliminated. Foreign investment is increasing, although the lifting of the US trade embargo had less effect than had been predicted. Despite its Communist government, Vietnam was admitted to the Association of South East Asian States in July 1995 because of its economic revival.

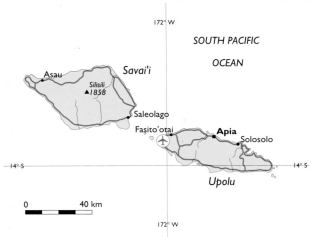

PHYSICAL GEOGRAPHY

Western Samoa is a group of seven small islands and two larger mountainous islands. Three-quarters of the population live on Upolu.

CLIMATE

The climate is tropical and very humid, cooler between May and November. There is heavy rainfall, mainly between December and April.

POLITICS AND RECENT HISTORY

Samoa is divided into two politically separate groups: Western Samoa, independent in 1962 from United Nations (UN) trusteeship and 40 years of New Zealand administration, and 100 km (60 miles) to the east the smaller American Samoa which is an external territory of the USA. Western Samoa still maintains close ties with New Zealand. A New Zealand High Commissioner is stationed in Apia and New Zealand acts as a channel of communication for Western Samoa in international affairs outside the Pacific region, through a treaty of friendship signed between the two coun-

tries in 1962.

Until recently, the Assembly of the Western Samoan government was traditionally elected by the *Matai*, or heads of families, but in the 1991 elections adult suffrage was introduced for the first time. The Human Rights Protection Party returned to power with a working majority. However, the Prime Minister Tofilau Eti Alesana has had to endure allegations of corruption and mismanagement within the national airline and his party may face a strong challenge from the recently formed Samoa Democracy Party at the next elections to parliament (the *Fono*).

The West Samoan Head of State for life is HH Malitoa Tanumafili II, but he has no real power.

THE INDEPEDENT STATE OF WESTERN SAMOA	
STATUS	Commonwealth State
AREA	2,840 sq km (1,095 sq miles)
POPULATION	163,000
CAPITAL	Apia
LANGUAGE	English, Samoan
RELIGION	Christian
CURRENCY	tala (dollar) (WST)
ORGANIZATIONS	Comm., UN

ECONOMY

The economy is primarily dependent on coconut products, coffee and bananas. In recent years the country has been stricken by cyclones – the most severe in 1990 and 1991. These devastated the coconut plantations. Consequently, the export of copra, important to the economy, has declined.

The country is self-sufficient in food products and there is some small light manufacturing industry producing items such as cigarettes and textiles. A small tourist trade is being encouraged, and Polynesian Airlines has introduced twice-weekly flights to Hawaii and a non-stop service to the USA.

Western Samoa has a trade deficit which has grown in the last few years, so that foreign aid, mainly from the UN and European Union countries, together with remittances from abroad, are vital. The nation is proud of its Polynesian culture but is very aware of the contrast between itself and its American Samoan neighbour which, with US support, has a per capita income three times as high.

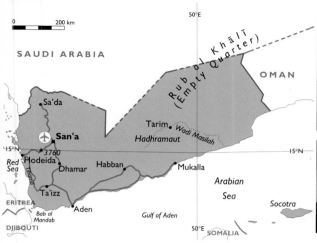

PHYSICAL GEOGRAPHY
Yemen is a land of mountains and desert. From the Red Sea, beyond the coastal plain known as Tihama, rises a rugged mountain range. Eastwards the land descends to the deserts of the Empty Quarter.

CLIMATE
In general Yemen has a hot and arid climate, although rainfall in the western mountains supports agriculture.

POLITICS AND RECENT HISTORY
North and South Yemen were unified in May 1990 under a provisional government. In April 1993 the first multi-party elections were held and, despite some allegations of corruption, appear to have been fair and peaceful. The People's General Congress (PGC), with its main following in the north, won most seats but not enough to form a government. It agreed to form a coalition with the other two leading parties, the Yemen Congregation for Reform, with Islamic credentials and support from Saudi Arabia, and the Yemeni Socialist Party based in the south. Ali Abdullah Saleh of the PGC became president.

In May 1994 unity collapsed as the south, under Vice-President Ali Salem al Beidh, sought to secede. Military defeat was inflicted by the north, who besieged the town of Aden and the United Nations was able to mediate and achieve a cease-fire. A fragile unity has been restored, but one in which the south distrusts and resents the authority of the north.

Although embroiled with internal problems, the Yemeni government is seeking to improve relations with Saudi Arabia and Kuwait. These were damaged because of Yemen's support for Iraq in the Gulf War. In a recent breakthrough, Yemen and Saudi Arabia were able to agree a formula for resolution of a long-standing border dispute.

THE REPUBLIC OF YEMEN	
STATUS	Republic
AREA	477,530 sq km (184,325 sq miles)
POPULATION	12,302,000
CAPITAL	San'a
LANGUAGE	Arabic
RELIGION	Sunni and Shi'a Muslim
CURRENCY	Yemeni dinar and rial
ORGANIZATIONS	Arab League, UN

ECONOMY
The merger between the two states was a combination of an essentially private sector economy in the north and a centrally planned Marxist system of state ownership in the south. This challenge has not yet been met, partly because of civil war an partly because of the manner in which tribal clans wield power.

Joint teams from the International Monetary Fund and the World Bank, who would be prepared to invest in the country, have been frustrated by the government's inability to carry through reform, in particular by cutting subsidies and curtailing expenditure on a large civil service.

There are, however, economic opportunities. Yemen is an oil producing nation and has recently concluded a deal with France to exploit natural gas resources. There are also plans to develop Aden, still an efficient port, to a status which its commanding position deserves. Success is dependent upon the resolve of the government to carry through reform and to achieve a real, rather than a nominal, unity.

YUGOSLAVIA (FEDERAL REPUBLIC)

PHYSICAL GEOGRAPHY
Montenegro, Kosovo and southern Serbia have rugged, mountainous, forested landscapes. Northern Serbia and Vojvodina are mainly low-lying, drained by the Danube system.

CLIMATE
The climate is continental with warm summers and cold winters. Rainfall is spread fairly evenly throughout the year, although snow is common in winter.

POLITICS AND RECENT HISTORY
Serbia and Montenegro, following the secession of other republics, are all that is left of former Yugoslavia. Its official title is the Federal Republic of Yugoslavia, including the Republics of Serbia and Montenegro.

President Slobodan Milosevic, the leader of the Socialist Party, was re-elected to the presidency of Serbia in December 1992. At the same time a coalition government led by the Socialists was formed, but in autumn 1993, in an apparent attempt to rid himself of the need for support from radicals, President Milosevic dissolved the government and called for new elections. They were held in December 1993 and, although the Socialists gained most seats, the need for a coalition was not eliminated.

Not long ago, all parties supported the concept of a Greater Serbia, to be achieved by uniting Serbian territories in Bosnia and Croatia. It is now clear that the will within Serbia to achieve this objective had diminished by the time the Dayton accord on Bosnia was signed in December 1995. Serbia made no move when Croatian forces attacked Serbs in Western Slavonia, and the Serbian government acted to persuade Bosnian Serbs to release UN hostages.

In any case, the President faces his own difficulties within Serbia's borders. The Albanian majority within Kosovo continue to press their demands for autonomy and eventual union with Albania, while Montenegro, irritated by the effects of sanctions, which it claims are not its responsibility, has threatened secession.

ECONOMY
By the end of 1993, sanctions had ruined the Serbian economy and caused severe damage to that of Montenegro. Hyperinflation ruled at levels exceeding those of Germany in the 1920s.

Since then there has been a recovery, largely due to some sound financial measures, the black market economy and widespread sanction breaking. The issue of new currency at parity with the Deutschmark brought stability and inflation was brought under control. Despite this improvement, the sanctions still hurt. Serbians are a great deal poorer than before the break-up of Yugoslavia and their situation will not improve for quite some time following the lifting of sanctions.

YUGOSLAVIA (FEDERAL REPUBLIC)	
STATUS	Federation of former Yugoslav Republics of Serbia and Montenegro
AREA	102,170 sq km (39,435 sq miles)
POPULATION	10,519,000
CAPITAL	Belgrade (Beograd)
LANGUAGE	Serbo-Croat, Albanian and Hungarian minorities
RELIGION	Orthodox Christian, 10% Muslim
CURRENCY	new dinar (YUD)
ORGANIZATIONS	UN (suspended)

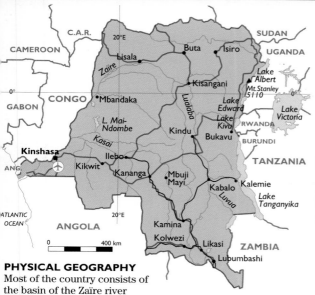

PHYSICAL GEOGRAPHY

Most of the country consists of the basin of the Zaïre river flanked by plateaux, with high mountain ranges to the north and east.

CLIMATE

The climate is tropical with rainforest close to the Equator and savannah in the northern and southern parts.

POLITICS AND RECENT HISTORY

The former Belgian Congo became independent in 1960 and in 1971 changed its name from Congo (Kinshasa) to Zaire.

Since 1965 Zaire has been ruled as a dictatorship by President Mobutu. In 1990 he agreed to multi-party elections and set up a transitional government (the High Council of the Republic) under Prime Minister Etienne Tshisekedi. In early 1993 Tshisekedi was replaced by Faustin Birindwa on the orders of the President, who still controlled the armed forces and the treasury. Tshisekedi, however, refused to leave office and continued to run government. Eventually, this bizarre state of affairs was

brought to an end when parliament appointed Kengo Wa Dondo as prime minister. He was able to command the support of both President Mobutu's factions and opposition elements. He has started to bring the country back from the brink of political anarchy and to root out corruption. His apparent ambition to start to tackle the massive economic problems has brought guarded encouragement from western nations and the possibility of renewed aid. Even so, he faces enormous political difficulties with some provinces threatening secession and other internal problems, such as the out-

break of Ebola fever in 1995.

ECONOMY

Zaire's economy has all but collapsed. It has massive foreign debts and has suffered astronomic inflation. The army and the civil service are infrequently paid and the country came close to expulsion from the International Monetary Fund.

The copper mines, once Zaire's main source of wealth, are closed and the only commodity providing revenue is diamonds.

Amid this chaos an informal economy is starting to flourish, but Prime Minister Kengo has a very difficult path to follow if he is to restore economic respectability.

THE REPUBLIC OF ZAIRE	
STATUS	Republic
AREA	2,345,410 sq km (905,330 sq miles)
POPULATION	41,231,000
CAPITAL	Kinshasa
LANGUAGE	French, Lingala, Kiswahili Tshiluba, Kikongo
RELIGION	traditional beliefs, 46% Roman Catholic, 28% Protestant
CURRENCY	zaire (ZRZ)
ORGANIZATIONS	OAU, UN

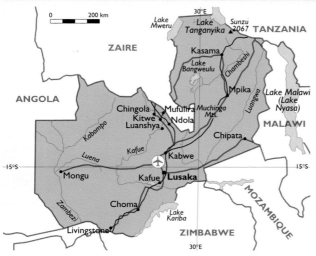

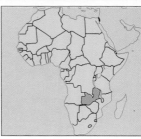

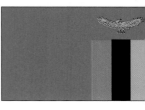

PHYSICAL GEOGRAPHY

The country is dominated by high savannah plateaux. Its southern boundary is the Zambezi River.

CLIMATE

The climate is tropical with three seasons: cool and dry, hot and dry, and then rainy between November and May.

POLITICS AND RECENT HISTORY

Formerly the British protectorate of Northern Rhodesia, then in 1953 part of the Federation of Rhodesia and Nyasaland, this state finally became independent as Zambia in 1964. The first head of government was the leader of the United National Independence Party, Dr Kenneth Kaunda, who declared a one-party state in 1972, when political conflict descended into violence.

In 1990 a referendum on a return to multi-party politics revealed support for an opposition group, the Movement for Multi-party Democracy (MMD). Multi-party elections were subsequently held and won by the MMD leader, Frederick Chiluba. Apart from the ruling MMD

party, the electorate now have a choice of two other well-supported parties, the UNIP and the recently formed National Party.

President Chiluba's government has been beset by accusations of corruption and, in particular, drug trafficking, which has plagued the country for over ten years. Due to international pressure, many of his ministers have been forced to resign or have been dismissed.

ECONOMY

President Chiluba is seeking to rebuild the Zambian economy by means of a structural readjustment programme. This will entail, among other measures, reversal of the state ownership policies of the previous administration and land reforms. The

big test will be the privatization of the giant Zambia Consolidated Copper Mines, which generates 90 per cent of Zambia's export revenue. In state hands it has become inefficient and is overmanned, although falls in copper prices have not helped.

Zambia is an agricultural nation although the potential is far from fully realized. Output has also suffered from the severe drought in 1992. Similarly Zambia has enormous tourist potential, but this resource is as yet hardly exploited. Investment is urgently needed in infrastructure and transportation systems if the economic opportunities are to be grasped. Certainly the country, with its low population and its natural riches, could become prosperous.

THE REPUBLIC OF ZAMBIA	
STATUS	Republic
AREA	752,615 sq km (290,510 sq miles)
POPULATION	8,936,000
CAPITAL	Lusaka
LANGUAGE	English, African languages
RELIGION	75% Christian, animist minority
CURRENCY	kwacha (ZMK)
ORGANIZATIONS	Comm., OAU, SADC, UN

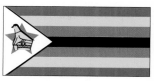

PHYSICAL GEOGRAPHY
Extensive high plateaux in central Zimbabwe are bordered by the Zambezi river valley with Lake Kariba in the north and the Limpopo in the south.

CLIMATE
Although the climate is tropical, temperatures are moderate because of the altitude of the high plateaux. Average rainfall is 740 mm (29 inches) a year, occurring mainly between November and March.

POLITICS AND RECENT HISTORY
When the Federation of Rhodesia and Nyasaland dissolved in 1963, Nyasaland became independent as Malawi, Northern Rhodesia as Zambia and Southern Rhodesia remained under British control. In defiance of Britain's pressure for Southern Rhodesia to accept black majority rule, Ian Smith's white government unilaterally declared independence, renaming the country Rhodesia. United Nations sanctions were imposed, but were withstood, and years of violence followed. In 1979 Rhodesia adopted majority rule and in 1980 Robert Mugabe became President of the new republic of Zimbabwe.

President Mugabe's ruling party, the Zimbabwe African National Union – Patriotic Front (ZANU–PF), has remained in power, having merged with Joshua Nkomo's ZANU party in 1989. In general elections in 1995, ZANU-PF won all but two of the contested seats, although the election was boycotted by the New Forum Party, led by Bishop Abel Muzorewa.

There were accusations of electoral mismanagement. Although President Mugabe maintains a powerful political grip on the country, his reputation has suffered. The most serious scandal was when land purchased from white farmers was allocated to ministers and senior party officials rather than, as intended, to black farmers. Although these allocations were eventually cancelled, the

THE REPUBLIC OF ZIMBABWE	
STATUS	Republic
AREA	390,310 sq km (150,660 sq miles)
POPULATION	10,739,000
CAPITAL	Harare
LANGUAGE	English, native languages
RELIGION	58% Christian, traditional beliefs
CURRENCY	Zimbabwe dollar (ZWD)
ORGANIZATIONS	Comm., OAU, SADC, UN

issue damaged the government's credibility.

ECONOMY
The Zimbabwean economy has suffered from mismanagement of public funds, although world recession and the disastrous drought of 1992 had a damaging effect.

Under pressure from the World Bank, President Mugabe has abandoned some of his socialist principles and embarked on a structural economic and adjustment programme. Subsidies on food have been removed, inflation has fallen and curbs on foreign investment have been lifted. It will, however, take time for beneficial effects of these reforms to be felt by ordinary Zimbabweans, especially as most of the nation's wealth is still in the hands of the minority white population.

Zimbabwe exports coffee, tea, fruit, vegetables and flowers. Tobacco, after several years decline, is now increasing in importance. Cotton, as well as being exported, is the basis for an indigenous textile industry. Nickel mining, in particular, has grown. Diamonds were first extracted in 1993 and the country's first platinum mine is being developed.

ANTARCTIC RESEARCH STATIONS
1 Teniente Rodolfo Marsh (Chile)
2 Comandante Ferraz (Brazil)
3 Capitán Arturo Prat (Chile)
4 Bellingshausen (Russian Fed.)
5 Teniente Jubany (Arg.)
6 Arctowski (Poland)
7 General Bernardo O'Higgins (Chile)
8 Esperanza (Arg.)
9 Vicecomodoro Marambio (Arg.)
10 Chang Cheng (Great Wall) (China)
11 Palmer (USA)
12 Vernadsky (Ukraine)
13 Rothera (UK)
14 Artigas (Urg.)
15 General San Martín (Arg.)

Scotia Sea

SOUTH ORKNEY IS. (UK)
Coronation I.
Signy (UK)
Orcadas (Arg)
Laurie I.
Clarence I.

SOUTH SHETLAND IS. (UK)
Elephant I.
14 2 10
3 1 5 6
4 5
7 8
9
11
12
13
15
16
Larsen Ice Shelf
GRAHAM LAND

Argentine Claim
British Antarctic Territory
Chilean Claim

Argentine Claim
British Antarctic Territory
Dronning Maud Land (Norway)

Chilean Claim

Antarctic Circle

60°
65°
0°

Weddell Sea

average minimum extent of sea

Lazarev Sea

Bellingshausen Sea

Ronne Ice Shelf
Korff Ice Rise
Berkner I.
Henry Ice Rise
Filchner Ice Shelf

General Belgrano II (Arg.)
Halley (UK)
Rüiser Larsenisen

Georg von Neumayer (Germany)
Sanae (S. Africa)
Fimbulisen

COATS LAND

75°

ELLSWORTH LAND
PALMER LAND
Antarctic
Peninsula

Vinson Massif
4897

Amundsen Sea

LESSER ANTARCTICA
MARIE BYRD LAND

Georg Forster (Germany)

Maitri (India)
Novolazarevskaya (Rus. Fed.)

Asuka (Japan)
Mt Victor
2588

85°

SOUTH POLE

Amundsen-Scott (USA)
Titan Dome
Pensacola Mts
Queen Maud Mts
Mt Kirkpatrick
4528

Showa (Japan)

45°

ENDERBY LAND

Molodezhnaya (Rus. Fed.)

Mawson (Aust.)

135°

KING EDWARD VII LAND
Roosevelt I.

Ross Ice Shelf
Crary Ice Rise

85°

Churchill Mts

South Geomagnetic Pole (1990)

Mt Menzies
3355

MAC. ROBERTSON LAND

Lambert Gl.
Amery Ice Shelf

McMurdo (USA)
Scott Base (NZ)
Mt Erebus
3794

Ross Sea

Ross Dependency (New Zealand)

GREATER ANTARCTICA

Transantarctic Mountains

VICTORIA LAND

Shafer Pk
3600

75°

WILHELM II LAND
PRINCESS ELIZABETH LAND

Zhongshan (China)
Davis (Aust.)

West Ice Shelf

Davis Sea

GEORGE V LAND
OATES LAND

QUEEN MARY LAND
Mt Amundsen
1380

Mirnyy (Rus. Fed.)

WILKES LAND

Shackleton Ice Shelf

180°

Cook Ice Shelf

Moscow Univ Ice Shelf
Casey (Aust.)

PACIFIC OCEAN

Dumont d'Urville (France)
65° Dumont d'Urville Sea

SOUTHERN OCEAN

90°

South Magnetic Pole (1990)

Australian Antarctic Territory

Terre Adélie (France)

Australian Antarctic Territory

INDIAN OCEAN

135°
55°

Abyssinia now Ethiopia **71**
Abkhazia autonomous region Georgia **79**
Aden Protectorate now Yemen **215**
Adjaria autonomous region Georgia **79**
Adygea republic Russian Federation **162**
Afars and Issas (French territory) now Djibouti **63**
Afghanistan 12
Alabama state USA **205**
Aland self-governing island province Finland **73**
Alaska state USA **205**
Albania 13
Alberta province Canada **44**
Alderney island Channel Islands **202**
Aleutian Islands territory USA **208**
Algeria 14
Altay republic Russian Federation **162**
American Samoa unincorporated territory USA **208**
American Virgin Islands external territory USA **208**
Andorra 15
Angola 16
Anguilla UK dependent territory **203**
Anhui (Anhwei) province China **50**
Anhwei see **Anhui**
Antarctica 220
Antigua and Barbuda 17
Argentina 18
Argentinian Antarctic claim 220
Arizona state USA **205**
Arkansas state USA **206**
Armenia 19
Aruba self-governing island Netherlands realm **141**
Ascension island dependency St Helena **203**
Ashmore and Cartier Islands external territory Australia **22**
Australia 20–2
Australian Antarctic Territory 220
Australian Capital Territory (Canberra) federal territory Australia **21**
Austria 23
Azerbaijan 24
Azores self-governing island region Portugal **158**
Baden-Württemberg state Germany **81**
Bahamas 25
Bahrain 26
Baleares see **Balearic Islands**
Balearic Islands (Baleares) island province Spain **178**
Banaba island Kiribati **106**
Bangladesh 27
Barbados 28
Bashkortostan republic Russian Federation **162**

Basutoland now Lesotho **114**
Bavaria see **Bayern**
Bayern (Bavaria) state Germany **81**
Bechuanaland Protectorate now Botswana **35**
Beijing (Peking) municipality China **50**
Belarus 29
Belau see **Palau**
Belgium 30
Belgian Congo now Zaire **217**
Belize 30
Belorussia see **Belarus**
Benin 32
Berlin state Germany **81**
Bermuda self-governing UK crown colony **202**
Bessarabia now Moldova **131**
Bhutan 32
Bijagós Archipelago islands of Guinea-Bissau **87**
Bioko island province Equatorial Guinea **69**
Bolivia 33
Bonaire self-governing island Netherlands Antilles, Netherlands **141**
Bonin Islands islands Japan **102**
Borneo island in Indonesia **94**; Malaysia **123**
Bosnia-Herzegovina 34
Botswana 35
Bougainville Island island Papua New Guinea **152**
Brandenburg state Germany **81**
Brazil 36
Bremen state Germany **81**
British Antarctic Territory 220
British Columbia province Canada **44**
British Guiana now Guyana **88**
British Honduras now Belize **31**
British Indian Ocean Territory UK dependency comprising the Chagos Archipelago **203**
British Solomon Islands now Solomon Islands **174**
British Somaliland now Somalia **175**
British Virgin Islands UK dependent territory **203**
Brunei 37
Bulgaria 38
Burkina 39
Burkina Faso see **Burkina**
Burma see **Myanmar**
Burundi 40
Buryatia republic Russian Federation **162**
Byelorussia see **Belarus**
California state USA **206**
Cambodia 41
Cameroon 42
Canada 43–4

Canary Islands island provinces Spain **179**
Canberra see **Australian Capital Territory**
Cape Verde 45
Cayman Islands UK dependent territory **203**
Celebes island Indonesia **94**
Central African Republic 46
Ceuta Spanish external province **179**
Ceylon now Sri Lanka **180**
Chad 47
Chagos Archipelago see **British Indian Ocean Territory**
Channel Islands British crown dependency **202**
Chechnia republic Russian Federation **162**
Chekiang see **Zhejiang**
Chile 48
Chilean Antarctic claim 220
China 49–51
Chinghai see **Qinghai**
Christmas Island external territory Australia **22**
Chuvash Republic Russian Federation **162**
Cocos (Keeling) Islands external territory Australia **22**
Colombia 52
Colorado state USA **206**
Comoros 53
Congo 54
Congo (Brazzaville) now Congo **54**
Congo (Kinshasa) now Zaire **217**
Connecticut state USA **206**
Cook Islands self-governing territory overseas in free association with New Zealand **143**
Coral Sea Islands external territory Australia **22**
Corsica island region France **74**
Costa Rica 55
Côte d'Ivoire 56
Crete island region Greece **83**
Crimea autonomous republic Ukraine **197**
Croatia 57
Cuba 58
Curaçao self-governing island Netherlands Antilles, Netherlands **141**
Cyprus 59
Czechoslovakia now Czech Republic **60**; Slovakia **174**
Czech Republic 60
Dagestan republic Russian Federation **162**
Dahomey now Benin **32**
Delaware state USA **206**
Denmark 61–2
District of Colombia federal district USA **206**
Djibouti 63

Dominica 64
Dominican Republic 65
Dronning Maud Land Norwegian Antarctic territory **220**
Dutch East Indies now Indonesia **94**
Dutch Guiana now Surinam **182**
East Pakistan now Bangladesh **27**
Eastern Cape province South Africa **176-7**
Eastern Transvaal now Mpumalanga **176-7**
Ecuador 66
Egypt 67
Eire see **Ireland**
Ellice Islands now Tuvalu **195**
El Salvador 68
England constituent country UK **201**
Equatorial Guinea 69
Eritrea 69
Estonia 70
Ethiopia 71
Falkland Islands UK crown colony **202**
Faroe Islands self-governing island region Denmark **62**
Fernando Poo now Bioko, Equatorial Guinea **69**
Fiji 72
Finland 73
Florida state USA **206**
Formosa now Taiwan **186**
France 74–6
Free State (Orange Free State) province South Africa **177**
French Antarctic Territory see **Terre Adélie**
French Guiana overseas department France **76**
French Guinea now Guinea **86**
French Polynesia overseas territory France **75**
French Somaliland now Djibouti **63**
French Sudan now Mali **125**
Fujian (Fukien) province China **50**
Fukien see **Fujian**
Gabon 77
Galapagos Islands archipelago province Ecuador **66**
Gambia 78
Gansu (Kansu) province China **50**
Gaza 98
Georgia 79
Georgia state USA **206**
Germany 80–1
Ghana 82
Gibraltar UK crown colony **202**
Gilbert Islands now Kiribati **106**

Gold Coast now Ghana **82**
Grande Comore island Comoros **53**
Great Britain see United Kingdom
Greece 83
Greenland self-governing island region Denmark **62**
Grenada 84
Grenadines islands see Grenada; St Vincent
Guadeloupe overseas department France **76**
Guam external territory USA **208**
Guangdong (Kwangtung) province China **50**
Guangxi-Zhuang (Kwangsi-Chuang) autonomous region China **50**
Guatemala 85
Guernsey island Channel Islands **202**
Guinea 86
Guinea-Bissau 87
Guizhou (Kweichow) province China **51**
Guyana 88
Hainan province China **51**
Haiti 89
Hamburg state Germany **81**
Hawaii state USA **206**
Heard and McDonald Islands external territory Australia **22**
Hebei (Hopei) province China **51**
Heilongjiang (Heilungkiang) province China **51**
Heilungkiang see Heilongjiang
Henan (Honan) province China **51**
Hesse see Hessen
Hessen (Hesse) state Germany **81**
Hispaniola islands comprising Haiti **89**; Dominican Republic **65**
Hokkaido island Japan **102**
Honan see Henan
Honduras 90
Hong Kong UK dependent territory (includes Kowloon and the New Territories) **203**
Honshu island Japan **102**
Hopei see Hebei
Hubei (Hupeh) province China **51**
Hunan province China **51**
Hungary 91
Hupeh see Hubei
Iceland 92
Idaho state USA **206**
Illinois state USA **206**
India 93
Indiana state USA **206**
Indonesia 94
Ingushetia republic Russian Federation **162**

Inner Mongolia see Nei Mongol
Iowa state USA **206**
Iran 95
Iraq 96
Ireland, Republic of 97
Irian Jaya province of Indonesia on New Guinea **94**
Isle of Man British crown dependency **202**
Israel 98
Italy 99
Ivory Coast see Côte d'Ivoire
Jamaica 100
Japan 101-2
Java (Jawa) island of Indonesia **94**
Jawa see Java
Jersey island Channel Islands **202**
Jiangsu (Kiangsu) province China **51**
Jiangxi (Kiangsi) province China **51**
Jilin (Kirin) province China **51**
Jordan 103
Kabardino-Balkar Republic Russian Federation **163**
Kalmykia-Khal'mg Tangch republic Russian Federation **163**
Kampuchea now Cambodia **41**
Kansas state USA **206**
Kansu see Gansu
Karachay-Cherkess Republic Russian Federation **163**
Karakalpak (Karakalpakstan) autonomous republic Uzbekistan **210**
Karakalpakstan see Karakalpak
Karelia republic Russian Federation **163**
Kazakhstan 104
Keeling Islands see Cocos Islands
Kentucky state USA **206**
Kenya 105
Khakassia republic Russian Federation **163**
Khmer Republic now Cambodia **41**
Kiangsi see Jiangxi
Kiangsu see Jiangsu
Kirghizia see Kyrgyzstan
Kiribati 106
Kirin see Jilin
Komi Republic Russian Federation **163**
Korea, North 107
Korea, South 108
Kowloon see Hong Kong
Kuwait 109
Kwangtung see Guangdong
Kwangsi-Chuang see Guangxi-Zhuang

Kwazulu-Natal province South Africa **176-7**
Kweichow see Guizhou
Kyrgyzstan 110
Kyushu island Japan **102**
Laos 111
Latvia 112
Lebanon 113
Lesotho 114
Liaoning province China **51**
Liberia 115
Libya 116
Liechtenstein 117
Lithuania 118
Louisiana state USA **206**
Lower Saxony see Niedersachsen
Luxembourg 119
Macao see Macau
Macau (Macao) Chinese territory under Portuguese administration **158**
Macedonia (former Yugoslav Republic) 120
Madagascar 121
Madeira self-governing island region Portugal **158**
Maine state USA **206**
Malagasy Republic now Madagascar **121**
Malawi 122
Malaya now Malaysia **123**
Malaysia 123
Maldives 124
Mali 125
Malta 126
Manitoba province Canada **44**
Mari El republic Russian Federation **163**
Marshall Islands 127
Martinique overseas department France **76**
Maryland state USA **206**
Massachusetts state USA **206**
Mauritania 128
Mauritius 129
Mayotte territorial collectivity France **75**
Mecklenburg-Vorpommern state Germany **81**
Melilla Spanish external province **179**
Mexico 130
Michigan state USA **206**
Micronesia 131
Minnesota state USA **206**
Mississippi state USA **206**
Missouri state USA **206**
Mohéli island Comoros **53**
Moldavia see Moldova
Moldova 131
Monaco 132
Mongolia 133
Montana state USA **206**
Montenegro republic Yugoslavia (Federal Republic) **216**
Montserrat UK crown colony **202**

Mordovian Republic Russian Federation **163**
Morocco 134
Mozambique 135
Muscat and Oman now Oman **148**
Myanmar 136
Nakhichevan autonomou republic Azerbaijan **24**
Namibia 137
Natal now Kwazulu-Natal **176-7**
Nauru 138
Nebraska state USA **207**
Nei Mongol (Inner Mongolia) autonomous region China **51**
Nepal 139
Netherlands 140–1
Netherlands Antilles self-governing part Netherlands Realm **141**
Nevada state USA **207**
New Brunswick province Canada **44**
New Caledonia overseas territory France **75**
Newfoundland province Canada **44**
New Guinea island comprising Irian Jaya, Indonesia **94**; part of Papua New Guinea **152**
New Hampshire state USA **207**
New Hebrides now Vanuatu **211**
New Jersey state USA **20**
New Mexico state USA **2**
New South Wales state Australia **21**
New Territories see Hong Kong
New York state USA **207**
New Zealand 142–3
New Zealand Antarctic Territory see Ross Dependency
Nicaragua 144
Niedersachsen (Lower Saxony) state Germany
Niger 145
Nigeria 146
Ninghsia Hui see Ningxia-Hui
Ningxia-Hui (Ninghsia Hui) autonomous regior China **51**
Niue self-governing territc overseas in free associatic with New Zealand **143**
Nordrhein-Westfalen sta Germany **81**
Norfolk Island external territory Australia **22**
North Carolina state USA **207**
North Dakota state USA **207**
Northern province South Africa **176-7**
Northern Cape province South Africa **176-7**
Northern Ireland constituent region UK **2**

Northern Mariana Islands self-governing commonwealth USA 208
Northern Rhodesia now Zambia 218
Northern Territory territory Australia 21
North Korea see Korea, North
North Ossetian Republic Russian Federation 163
North West province South Africa 176-7
Northwest Territories territory Canada 44
North Yemen now Yemen 215
Norway 147
Norwegian Antarctic Territory see Dronning Maud Land
Nova Scotia province Canada 44
Nyasaland now Malawi 122
Ocean Island see Banaba
Ohio state USA 207
Oklahoma state USA 207
Oman 148
Ontario province Canada 44
Orange Free State see Free State
Oregon state USA 207
Pakistan 149
Palau (Belau) 150
Panama 151
Papua New Guinea 152
Paraguay 153
Peking see Beijing
Pennsylvania state USA 207
Peru 154
Philippines 155
Pitcairn Islands UK dependent territory 203
Poland 156
Portugal 157-8
Portuguese Guinea now Guinea-Bissau 87
Prince Edward Island province Canada 44
Puerto Rico self-governing commonwealth USA 208
Qatar 159
Qinghai (Chinghai) province China 51
Québec province Canada 44
Queensland state Australia 21
Réunion overseas department France 76
Rheinland-Pfalz state Germany 81
Rhode Island state USA 207
Rhodesia now Zimbabwe 219
Romania 160
Ross Dependency New Zealand Antarctic Territory 220
Russian Federation 161-3
Ruanda-Urundi now Burundi 40; Rwanda 164
Rwanda 164
Saarland state Germany 81

Sachsen (Saxony) state Germany 81
Sachsen-Anhalt state Germany 81
St Kitts and Nevis 164
St Helena UK dependent territory 203
St Lucia 165
St Pierre & Miquelon territorial collectivity France 75
St Vincent 165
Saba self-governing island Netherlands Antilles, Netherlands 141
Sakha republic Russian Federation 163
Samoa see American Samoa 208; Western Samoa 214
San Marino 166
São Tomé and Príncipe 166
Sardinia island region Italy 99
Sark island Channel Islands 202
Saskatchewan province Canada 44
Saudi Arabia 167
Saxony see Sachsen
Schleswig-Holstein state Germany 81
Scotland constituent country UK 201
Senegal 168
Serbia republic Yugoslavia (Federal Republic) 216
Seychelles 169
Shaanxi (Shensi) province China 51
Shandong (Shantung) province China 51
Shanghai municipality China 51
Shansi see Shanxi
Shantung see Shandong
Shanxi (Shansi) province China 51
Shensi see Shaanxi
Shikoku island Japan 102
Siam now Thailand 189
Sichuan (Szechwan) province China 51
Sierra Leone 170
Singapore 171
Sinkiang Uighur see Xinjiang Uygur
Sint Eustatius self-governing island Netherlands Antilles, Netherlands 141
Sint Maarten self-governing island Netherlands Antilles, Netherlands 141
Slovakia 172
Slovenia 173
Solomon Islands 174
Somalia 175
South Africa 176-7
South Australia state Australia 21
South Carolina state USA 207

South Dakota state USA 207
South Georgia and the South Sandwich Islands UK dependent territory 203
Southern Rhodesia now Zimbabwe 219
South Korea see Korea, South
South Ossetia region Georgia 79
Southwest Africa now Namibia 137
South Yemen now Yemen 215
Spain 178-9
Spanish Guinea now Equatorial Guinea 69
Spanish North Africa see Ceuta 179; Melilla 179
Sri Lanka 180
Sudan 181
Surinam 182
Swaziland 182
Sweden 183
Switzerland 184
Syria 185
Szechwan see Sichuan
Tadzhikistan see Tajikistan
Taiwan 186
Tajikistan 187
Tanganyika see Tanzania
Tanzania 188
Tasmania state Australia 21
Tatarstan republic Russian Federation 163
Tennessee state USA 207
Terre Adélie French Antarctic territory 220
Texas state USA 207
Thailand 189
Thüringen (Thuringia) state Germany 81
Thuringia see Thüringen
Tianjin (Tientsin) municipality China 51
Tibet see Xizang Zizhiqu
Tientsin see Tianjin
Timor island of Indonesia 94
Togo 190
Tokelau territory overseas New Zealand 143
Tonga 190
Trinidad and Tobago 191
Trucial States now United Arab Emirates 198
Tunisia 192
Turkey 193
Turkmenistan 194
Turks and Caicos Islands UK dependent territory 203
Tuva republic Russian Federation 163
Tuvalu 195
Udmurt Republic Russian Federation 163
Uganda 196
Ukraine 197
United Arab Emirates 198
United Kingdom 199-203
United States of America 204-8

Upper Volta now Burkina 39
Uruguay 209
USSR now Armenia 19; Azerbaijan 24; Belarus 29; Estonia 70; Georgia 79; Kazakhstan 104; Kyrgyzstan 110; Latvia 112; Lithuania 118; Moldova 131; Russian Federation 161-3; Tajikistan 187; Turkmenistan 194; Ukraine 197; Uzbekistan 210
Utah state USA 207
Uzbekistan 210
Vanuatu 211
Vatican City 211
Venezuela 212
Vermont state USA 207
Victoria state Australia 21
Vietnam 213
Virginia state USA 207
Virgin Islands (UK) see British Virgin Islands
Virgin Islands (USA) see American Virgin Islands
Wales principality UK 201
Wallis and Futuna Islands self-governing overseas territory France 75
Washington state USA 207
West Bank 98
Western Australia state Australia 21
Western Cape province South Africa 176-7
Western Sahara territory in dispute 134
Western Samoa 214
West Pakistan see Pakistan
West Virginia state USA 207
Wisconsin state USA 207
Wyoming state USA 207
Xinjiang Uygur (Sinkiang Uighur) autonomous region China 51
Xizang Zizhiqu (Tibet) autonomous region China 51
Yakutia see Sakha
Yemen 215
Yugoslavia now Bosnia-Herzegovina 34; Croatia 57; Macedonia 120; Slovenia 173; Yugoslavia (Federal Republic) 216
Yugoslavia (Federal Republic) 216
Yukon Territory territory Canada 44
Yunnan province China 51
Zaire 217
Zambia 218
Zanzibar island Tanzania 188
Zhejiang (Chekiang) province China 51
Zimbabwe 219